Introduction to

Relativistic Heavy Ion Physics

Introduction to

Relativistic Heavy Ion Physics

J. Bartke

Institute of Nuclear Physics
Cracow, Poland

World Scientific

NEW JERSEY • LONDON • SINGAPORE • BEIJING • SHANGHAI • HONG KONG • TAIPEI • CHENNAI

Published by

World Scientific Publishing Co. Pte. Ltd.
5 Toh Tuck Link, Singapore 596224
USA office: 27 Warren Street, Suite 401-402, Hackensack, NJ 07601
UK office: 57 Shelton Street, Covent Garden, London WC2H 9HE

British Library Cataloguing-in-Publication Data
A catalogue record for this book is available from the British Library.

INTRODUCTION TO RELATIVISTIC HEAVY ION PHYSICS

ISBN-13 978-981-02-1231-5
ISBN-10 981-02-1231-3

Printed in Singapore.

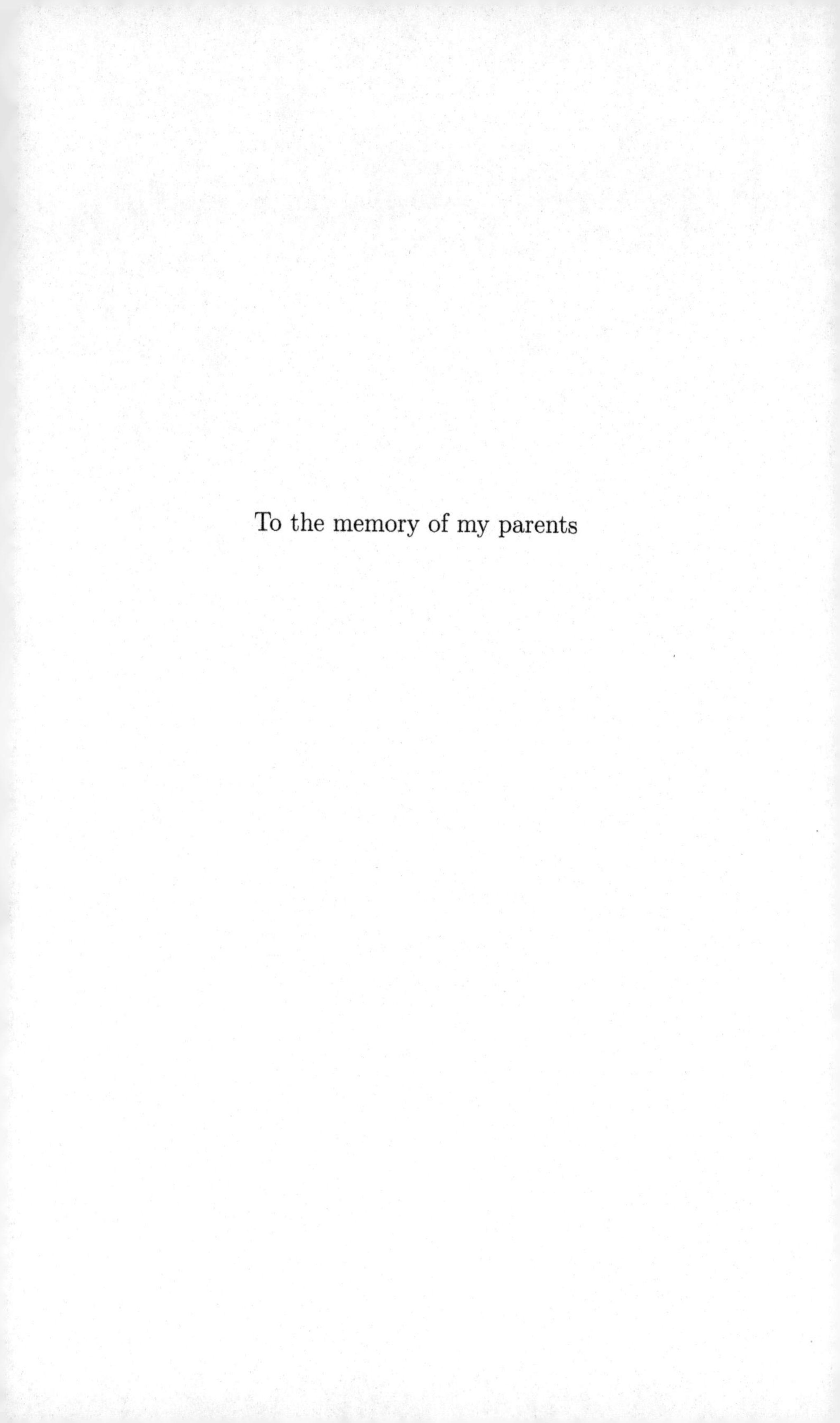

To the memory of my parents

Preface

The book is based on lectures for students specializing in high energy physics delivered by the author in Cracow over several years. It aims to present the physics of relativistic heavy ions, also called the relativistic nuclear physics. This field, only some thirty years old, is a highly interesting one, as collisions of relativistic heavy ions are believed to lead to the formation of a new state of matter — the quark-gluon plasma (QGP), thus bridging frontiers between nuclear and hadronic physics.

Only well established experimental results are presented in this book. The plots shown are not necessarily the most recent ones, they have been selected for their clarity, and for presenting well the main features of experimental data. Some recent results bearing large errors which do not show the trend of the data in an unambiguous way have not been included. For the same reason, in some cases plots carrying the label "preliminary" have been used if the "final" version could not be found, or if the new presentation of the same data was not so clear, e.g. data points were obscured by some curves resulting from theoretical calculations which we do not pretend to discuss in details. It is commonly known that in most cases the "final" corrections and refinements do not change trends of the data in an appreciable way.

On the theory side, apart from a chapter devoted to Quantum Chromodynamics (QCD), only simple theoretical ideas and phenomenological models which are well supported by experimental data are briefly discussed, without going too deep into the underlying mathematical formalism. Much care has been taken in the clarity of the presentation, to make the experimental results understandable using simple conjectures, and to show connection between various aspects of the data. Some recent ideas concerning the quark-gluon plasma, and a selection of predictions for nuclear collisions expected soon in the Large Hadron Collider at CERN complete the book.

We hope that the book will be appreciated by students and young researchers becoming involved in physics of relativistic nuclei, as well as by those looking for a comprehensive and up-to-date review of the field.

Jerzy Bartke
Professor of Physics
Institute of Nuclear Physics PAN
Cracow, Poland
August 2008

Acknowledgments

Special thanks are due to the author's wife Jadwiga for her constant encouragement and patience.

The book would not have taken shape without the invaluable technical help of Mrs Danuta Filipiak.

Contents

Chapter 1

Introduction

Relativistic heavy ion physics is a fascinating field. In a collision of two nuclei occuring at very high energy, whether in a fixed-target or in a collider mode, thousands of new particles are produced. Their identity and kinematical characteristics go beyond that what could be expected from a simple superposition of elementary nucleon-nucleon collisions, indicating the presence of some new phenomena.

Single events of high energy nuclear collisions were studied already in the 1950s in nuclear emulsions irradiated by cosmic rays in stratospheric balloon flights. It was found that the produced particles are strongly forward/backward collimated, with transverse momentum components limited to the values of the order of only a few hundred MeV/c. These early observations led to the notions of "multiple particle production", and of "fireballs", and stimulated new theoretical ideas. We shall limit ourselves to quote here just three papers of basic importance to this field.

Enrico Fermi [1] proposed a description of high energy hadronic (and nuclear) collisions in terms of the statistical thermal model, assuming a formation of a highly excited intermediate state, *a little fireball*, in which a thermal equilibration is reached, and the decay into final state particles follows the statistical rules.

Lev D. Landau [2] proposed a model in which the energy deposition in a small volume, of the size of the Lorentz-contracted nuclei, leads to the formation of a transient state which then undergoes a hydrodynamical expansion. While expanding, the system cools down until it reaches the freeze-out temperature T_f, being of the order of the pion mass, at which the formed hadrons become free particles.

Later, Rolf Hagedorn [3], while studying the mass spectrum of the then recently discovered numerous hadronic resonant states, made a conjecture that a multi-hadron state should be described by thermodynamics with a limiting temperature. This temperature, now quoted as the *Hagedorn temperature*, turned out to be about 160 MeV, or again of the order of the pion mass.

Systematic experimental studies of collisions of relativistic nuclei in laboratory conditions began in the early 1970-ies, when at the Lawrence Laboratory in Berkeley, USA, and at the Joint Institute of Nuclear Research in Dubna, Russia, light nuclei were accelerated to energies of a few GeV per nucleon, using the old proton synchrotrons.

At the beginning, these experiments seemed not to promise anything exciting, but soon a hypothesis was formulated that at high temperatures and densities the hadronic matter should undergo a phase transition to a state of free quarks and gluons, called *the quark-gluon plasma* (QGP) [4–6]. In the following years the Quantum Chromodynamics (QCD) was developed as the theory of strong interactions, and calculations on the lattice (LQCD) led to more precise predictions. A phase diagram for the strongly interacting matter with the phase boundary between hadronic matter and quark-gluon plasma was drawn, and critical values of the temperature and density were determined to be $T_c \cong$ (150–170) MeV, surprisingly close to the Hagedorn temperature, and $\rho_c \cong$ (1–2) GeV/fm^3, about ten times the matter density in nuclei. Conservative estimates show that such values could be reached in collisions of relativistic nuclei.

As quark-gluon plasma is believed to be the state of matter which existed for some microseconds after the Big Bang, collisions of relativistic heavy ions should recreate the conditions of the Early Universe. With this exciting hypothesis in mind, experiments with relativistic heavy ions became extremely interesting, and the program of accelerating heavy ions to much higher energies was launched at Brookhaven National Laboratory, USA, and at CERN, Geneva, Switzerland. At BNL ions were accelerated in the Alternating Gradient Synchrotron (AGS) to energies $\sim$ 10 GeV/nucleon, rather too low for the QGP search, but detectors and analysis methods suitable for studying multi-particle events were developed, and useful experience gained. At CERN ions were accelerated in the Super Proton Synchrotron (SPS) to energies ranging from 200 GeV/nucleon for light ions to 158 GeV/nucleon for those of lead. In the year 2000 the Relativistic Heavy Ion Collider RHIC was put into operation at BNL, increasing the effective collision energy (i.e. the energy in the centre-of-mass system) by another factor of ten.

Figure 1.1 shows a simplified picture of a central collision of two highly

Fig. 1.1 Simplified picture of a central collision of two high energy nuclei in the centre-of-mass frame. Colliding nuclei appear as thin discs due to Lorentz contraction.

relativistic nuclei in the centre-of-mass reference frame. The colliding nuclei are Lorentz-contracted, and thus appear as thin discs. In the central region, where the energy density is the highest, a new state of matter — the quark-gluon plasma — is supposedly created. The plasma expands and cools down,[a] quarks combine into hadrons and their mutual interactions cease when the system reaches the *freeze-out temperature.* A multi-hadron final state is formed, and free hadrons move towards the detectors.

Figure 1.2 shows the space-time evolution of a collision process, plotted in the light-cone variables (z, t). The two highly relativistic nuclei, identified in the Figure as "projectile" and "target", move essentially along the light cone, until they collide in the centre of the diagram. Nuclear fragments emerge from the collision again along the (forward) light cone, while the matter between the fragmentation zones populates the central region. This hot and dense matter is believed to be in the state of QGP. Interactions within it bring the system into local statistical equilibrium, and its further evolution can be described by relativistic hydrodynamics.[b] The surfaces of constant proper time, delineating various stages of this evolution, are approximately hyperbolae in this representation, as shown in the figure. The hydrodynamic description of high energy nuclear collisions was developed in many subsequent papers [8–11], see also reviews [12, 13]. Dynamical particle producing reactions, described with dissipative and diffusion terms, have

[a] According to recent theoretical ideas, the system may pass through some intermediate states with different properties.

[b] The problem is that in order to reproduce correctly the experimental results, the hydrodynamic evolution should start at the time below 1 fm/c after the collision, what means a very short equilibration time [14, 15]. The hypothesis of an *instantaneous thermalization* was, however, discussed already more than 20 years ago — see *e.g.* Ref. [11].

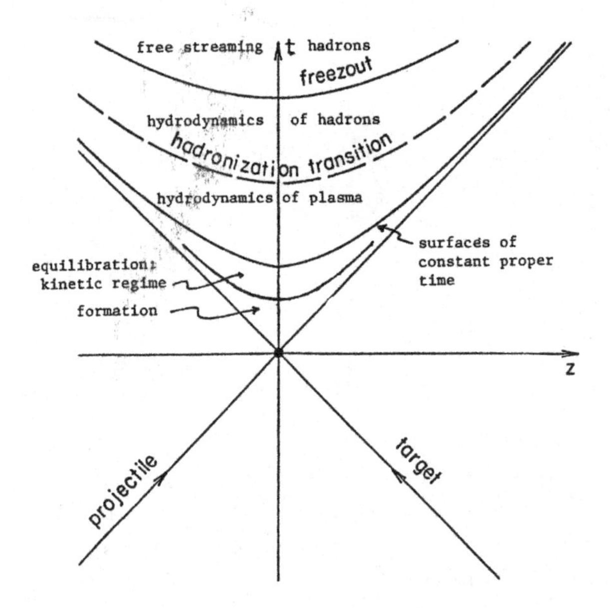

Fig. 1.2 Space-time evolution of a collision process of ultrarelativistic nuclei plotted in light-cone variables (from Ref. [7]).

been incorporated into relativistic hydrodynamics in Ref. [16], and the case of strangeness and/or heavy quarks has been discussed in Ref. [17].

Let us now try to estimate the energy density reached in relativistic heavy ion collisions. An estimate of the initial energy density is not straightforward: taking that of a Lorentz-contracted nuclei leads to unreasonably high values, and the so-called *Bjorken formula* is being used for this purpose. Bjorken [8] developed a rapidity-independent version of the Landau's hydrodynamical model in which the created transverse energy density, dE_T/dy, is related to the initial energy density, ϵ, by the formula

$$\epsilon = \frac{1}{\tau_f \, S} \frac{dE_T}{dy} \cong \frac{3}{2} \frac{\langle m_T \rangle}{\tau_f \, S} \frac{dN_{\rm ch}}{dy} \tag{1.1}$$

where τ_f is the *formation time*, conventionally taken to be $\tau_f = 1$ fm/c,[c] S is the transverse overlap area of the colliding nuclei (for a central collision of two identical nuclei of radius R this is simply $S = \pi R^2$), $\langle m_T \rangle$ is the

[c]It should be pointed out that this choice of the formation time is arbitrary. Intuitively, the formation time should be at least as long as the "crossing time" of the colliding nuclei, what means $\tau_f \geq 2R/\gamma$ where γ is the Lorentz factor of the colliding nuclei. For $\tau_f = 1$ fm/c this condition becomes to be valid at SPS energies.

mean value of the transverse mass of secondary particles, and $dN_{\rm ch}/dy$ is the measured density of charged secondary particles per unit of rapidity. The approximate relation $N \cong \frac{3}{2}N_{\rm ch}$ is assumed. The values of the energy density obtained from this formula vary between the following limits

$$(2\text{–}3)\ \mathrm{GeV/fm^3} \leq \epsilon \leq (5\text{–}6)\ \mathrm{GeV/fm^3}\ ,$$

the lower values corresponding to energies reached in the CERN SPS, and the higher ones to those of RHIC. Let us note that already the lower values exceed the critical density for the phase transition obtained from LQCD. This, together with the temperature $T \approx 140$ MeV obtained from the secondary particle spectra, and with the observation of some other phenomena predicted by theorists as signatures of the phase transition, led CERN to announce in February 2000 the discovery of the quark-gluon plasma.

Investigations of collisions of relativistic heavy ions have been further carried out at RHIC. Some new features have been observed, in particular a substantial collective flow was found in the emission pattern of secondary hadrons, what means that the created hot and dense system is rather a liquid than a gas. This "liquid" has a small viscosity, its properties are close to those of the "perfect" liquid. The *quark number scaling*, observed when comparing the flow of different particle species, points towards a partonic intermediate state, and can be considered as a strong evidence for the quark-gluon plasma.

The Large Hadron Collider (LHC) which should come into operation at CERN in the nearest time, will offer possibilities for investigations of collisions of heavy ions (Pb+Pb) at much higher energies. One can expect a substantial increase of the created energy densities, perhaps up to 10 GeV/fm^3, or even higher, and a longer lifetime of the created system.

References

[1] E. Fermi, *Progr. Theor. Phys.* **5** (1951) 570; *Phys. Rev.* **81** (1951) 683.
[2] L. D. Landau, *Izv. Akad. Nauk, Ser. Fiz.* **1** (1953) 51.
[3] R. Hagedorn and J. Rafelski, *Phys. Lett. B* **97** (1980) 136; R. Hagedorn, *Z. Phys. C* **17** (1983) 265, and earlier references therein.
[4] J. C. Collins and M. J. Perry, *Phys. Rev. Lett.* **34** (1975) 1353.
[5] N. Cabibbo and G. Parisi, *Phys. Lett. B* **59** (1975) 67.
[6] E. V. Shuryak, *Phys. Lett. B* **78** (1978) 150.
[7] G. Baym, *Proc. Quark Matter'84*, ed. K. Kajantie, Lecture Notes in Physics, Vol. 221 (Springer, 1985).
[8] J. D. Bjorken, *Phys. Rev. D* **27** (1983) 140.

[9] K. Kajantie and L. McLerran, *Nucl. Phys. B* **214** (1983) 261.
[10] G. Baym, *Nucl. Phys. A* **418** (1984) 525c; *Phys. Lett. B* **138** (1984) 18.
[11] R. Raitio, *Nucl. Phys. A* **418** (1984) 539c.
[12] P. Huovinen, *Quark Gluon Plasma 3*, eds. R. Hwa and X.-N. Wang (World Scientific, 2004), p. 600.
[13] P. F. Kolb and U. Heinz, *Quark Gluon Plasma 3*, eds. R. Hwa and X.-N. Wang (World Scientific, 2004), p. 634.
[14] U. Heinz, hep-ph/0204061.
[15] A. Białas, M. Chojnacki and W. Florkowski, *J. Phys. G: Nucl. Part. Phys.* **35** (2008) 104073.
[16] H. Th. Elze, J. Rafelski and L. Turko, *Phys. Lett. B* **506** (2001) 123.
[17] L. Turko, *J. Phys. G: Nucl. Part. Phys.* **35** (2008) 044919.

Chapter 2

Quantum Chromodynamics and the Phase Transition in Strongly Interacting Matter

Quantum chromodynamics (QCD) is the present-day theory of strong interactions. It was formulated in the years 1972–73 by Murray Gell-Mann and Steven Weinberg, both Nobel Prize winners. Its name, analogous to that of quantum electrodynamics (QED), the theory of electromagnetic interactions in which the electric charge plays a key role, refers to *colour charge* characterising the fundamental particles of strong interactions: quarks q and gluons g. With six quark flavours (u, d, s, c, b, t) and three colours, one can build all known "elementary" hadrons according to the simple rule: mesons are composed of quark and antiquark, $M = q_i\bar{q}_j$, and baryons are composed of three quarks, $B = q_iq_jq_k$, with indices i, j, k =1,2,...6. The corresponding antiparticles will then be $\bar{M} = \bar{q}_iq_j$ and $\bar{B} = \bar{q}_i\bar{q}_j\bar{q}_k$. QCD is based on the SU(3) symmetry group which gives the correct decomposition of the products of the basic triplet representations, with only $q\bar{q}$ and qqq configurations.

With the success in hadron classification it is, however, difficult to perform dynamical calculations in QCD using methods developed for QED, as in contrast to the small value of the electromagnetic coupling constant $\alpha = e^2/\hbar c = 1/137$, the QCD coupling constant α_s might be of the order of one. In fact, α_s is not constant, but it depends on the momentum Q^2 transferred in the interaction and shows a logarithmic decrease with increasing Q^2

$$\alpha_s(Q^2) \propto \frac{1}{\ln(Q^2/\Lambda^2)} \tag{2.1}$$

with Λ - a constant defining the scale. Thus for large values of Q^2 ("hard" collisions) we enter the perturbative region and can use calculation methods developed for QED, while for small Q^2 ("soft" collisions) one can only try to use very complicated numerical methods which will be discussed later in this chapter. Figure 2.1 shows the QCD coupling constant as a function of the distance which is the variable "inverse" to Q^2 (large Q^2 probe small distances). The vertical band represents the region intermediate between the "perturbative QCD" at small distances and the "strong QCD" at distances close to the nucleon radius.

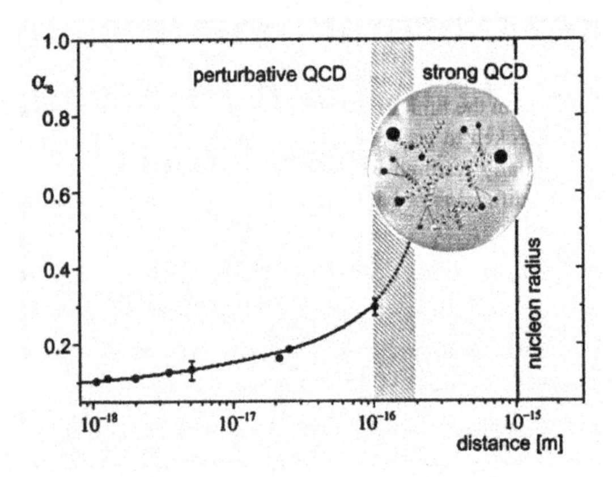

Fig. 2.1 Coupling constant of the strong interaction as function of distance. Experimental values are shown with small points (from Ref. [1]).

The QCD potential between a quark and an antiquark has, besides the obvious Coulomb term, another term which is linear in the separation distance, r:

$$V(r) = -\frac{\alpha}{r} + \sigma r \tag{2.2}$$

This latter term resembles the potential of the rubber band, and thus an intuitive picture is that an elastic string, or colour flux tube, forms between the quark and the antiquark. Its "elastic" properties are determined by the constant σ. With increasing mutual distance r, the attractive force increases, so as quarks cannot be separated. This is called *confinement* and experimentally it means that one cannot observe a free quark — all quarks are bound within hadrons. Moreover, one expects that at a point when energy of the stretched string becomes bigger than two quark masses, the string would break and a new quark-antiquark pair be created from the vacuum at the breaking point, forming a meson. This mechanism is supposed to be the source of hadron production in *string models* of high energy hadronic interactions.

Another chararacteristic feature of QCD, besides the quark confinement, is the *asymptotic freedom* meaning that at very short distances quarks behave as free particles. A possibility of quark liberation was noticed already in 1975 by Collins and Perry [2], who wrote: 'the quark model implies that superdense matter consists of quarks rather than of hadrons'. A new phase of matter in which quarks are no longer confined was considered a few months later by Cabibbo and Parisi [3]. The notion of *quark-gluon plasma* (QGP) was introduced in 1978 by Shuryak [4]. At high temperatures and/or densities a phase transition from hadronic matter to QGP should take place. Characteristics of this new phase of strongly interacting matter can be obtained from QCD only by numerical simulations as they are related to low-energy, non-perturbative properties of the theory.

In order to see similarities and differences between QCD and QED, and to visualize the much more complicated structure of QCD, we shall write some basic formulae for both theories.

The lagrangian in QED is

$$\mathcal{L} = -\frac{1}{4} F_{\mu\nu}^2 \tag{2.3}$$

with

$$F_{\mu\nu} = \partial_\mu A_\nu - \partial_\nu A_\mu \tag{2.4}$$

where A_μ is the vector potential of the electromagnetic (photon) field. The antisymmetric 4×4 tensor $F_{\mu\nu}$ contains components of the electric and magnetic fields.

The lagrangian in QCD with gluons only is

$$\mathcal{L}_o = -\frac{1}{4} \sum_i F_{\mu\nu}^a F^{\mu\nu\, a} \tag{2.5}$$

with $a = 1, 2, \ldots, 8$ colour indices, and

$$F_{\mu\nu}^a = \partial_\mu A_\nu^a - \partial_\nu A_\mu^a + g_s [A_\mu, A_\nu]^a \tag{2.6}$$

where A_μ^a are the eight vector potentials of the gluon field, and g_s is the strong (colour) charge. The last term in square brackets represents self-interaction of gluons, due to their non-zero charge (there is no similar term in QED as photons carry no charge).
If quarks are also present, the lagrangian acquires additional terms which describe interactions of quarks and gluons, and the self-interaction of quarks

$$\mathcal{L} = \mathcal{L}_o + i \sum_q \bar{\psi}_q^i \gamma^\mu (D_\mu)_{ij} \psi_q^j - \sum_q m_q \bar{\psi}_q^i \psi_{q\, i} \tag{2.7}$$

with

$$(D_\mu)_{ij} = \delta_{ij} \partial_\mu + i \frac{g_s}{2} \sum_a \lambda_{i,j}^a A_\mu^a \tag{2.8}$$

where ψ_q^i are the four-component Dirac spinors describing quark fields of colour i and flavour q and the (3×3) matrices $\lambda_{i,j}$ are the representations of the SU(3) group. The coupling constant α_s is related to the colour charge

$$\alpha_s = g_s^2/4\pi \tag{2.9}$$

The full expression for α_s is

$$\alpha_s(Q^2) = \frac{4\pi}{\beta_0 \ln(Q^2/\Lambda^2)}(1 + \text{higher logarithmic terms}) \tag{2.10}$$

with $\beta_0 = 11 - \frac{2}{3}n_f$ where n_f is the number of light quarks. According to the generally adopted convention one defines the value of α_s at the mass of the Z^0 boson to be $\alpha(m_Z) = 0.118 \pm 0.002$. This gives for the scale constant the value $\Lambda = 217^{+25}_{-23}$ MeV [5].

Large number of the field components, and the presence of the self-interaction terms, together with a large value of the QCD coupling constant α_s, make the calculations of the non-perturbative QCD very complicated. The numerical calculations employ the *lattice* approach, introduced in 1974 by Kenneth Wilson (Nobel Prize 1982). The computation consists in generating quark and gluon field configurations, weighted by the Boltzmann factor $\exp(-S)/Z$, where S is the action defined as the four-dimensional integral of the lagrangian density

$$S = \int d^4x\, \mathcal{L}(x) \tag{2.11}$$

and Z is the partition function for the system components, and then calculating the expectation value of the operator relevant to the problem studied, over these configurations. Numerical integration is performed on the four-dimensional lattice in (x, y, z, t) space. A typical lattice size is a few fm, and point spacing is of the order of 0.1 fm, with usually less points in the t-direction. As in any numerical integration, the precision increases with increasing number of points on the lattice. The maximum size of the lattice is determined by the present-day computing possibilities. Recently, some results have been obtained with the $32^3 \times 16$ lattice, and lattices as large as $64^3 \times 24$ are being investigated.

The procedure of the Monte-Carlo calculation is the following. One creates on the lattice a configuration of gluon and quark fields in which the numbers of quarks and antiquarks in each node are the same (with several types of quarks each type of them should be balanced independently). The action S is calculated, following the Wilson's prescription, as the sum of contour integrals over all *plaquettes* (elementary squares) of the lattice

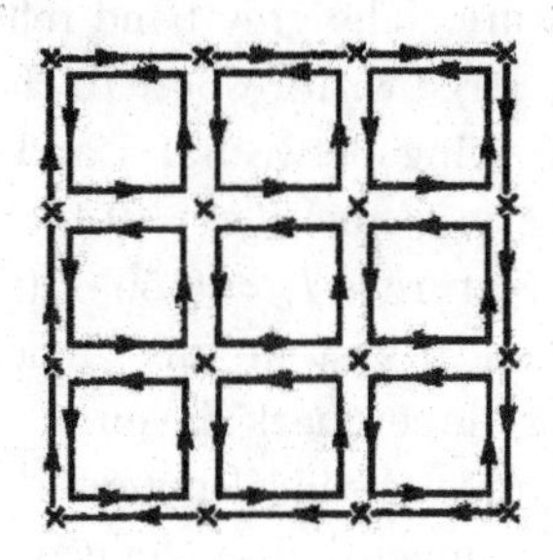

Fig. 2.2 The 3×3 Wilson loop covered with plaquettes.

(Fig. 2.2). Then the values of the field are randomly changed, with configurations yielding lower values of S (or larger Boltzmann factor $\exp(-S)$) retained. This procedure, based on the famous Metropolis algorithm, drives the system towards the equilibrated state, allowing to reveal phase transitions eventually occuring on this way. Physics results are obtained in the continuum limit, with the lattice constant going to zero.

In order to get a feeling why for the lattice QCD (LQCD) calculations a very high computer power is required, one should keep in mind that a four-dimensional cubic lattice of size N has N^4 nodes and $6\,N^4$ plaquettes, the gluon field has eight components, and in the SU(3) theory one is dealing with 3×3 complex matrices, each having 18 real parameters. With three quark flavours there are $4\times18 = 72$ real variables in each node. Lattice QCD pioneered the use of computers with parallel architecture. Multiple processors are arranged in a 3-dimensional array, the system also has a distributed memory. Actually both commercial machines and purpose-built machines employing customized processors are being used. The most widely used commercial machine is Cray/SGI, varying in size from few hundred to 1024 processors, with a peak speed approaching 1 TFlop (10^{12} floating point operations per second). Special computer systems containg up to about 10,000 processors have also been built, attaining a similar performance. In order to achieve better accuracy in the calculations, the LQCD groups are aiming at multi-TFlops computers in the years to come.

As it has already been pointed out earlier, from such properties of QCD as quark structure of hadrons and asymptotic freedom, the deconfined phase, or quark-gluon plasma (QGP), should exist at high temperatures and/or high energy densities. Figure 2.3 shows a corresponding schematic phase diagram, as obtained in earlier calculations with massless quarks, and often shown in literature. The grey band reflects the uncertainties of the calculation, especially large with respect to the critical energy density, the estimated value for it lying between 1.5 and 2.0 GeV/fm^3, or about ten times the matter density in nuclei $\rho_0 = 0.16$ GeV/fm^3. The predicted value for the critical temperature is $T_c = (150\text{–}170)$ MeV, depending on the assumptions about the studied system (the higher value is for two quark flavours, the lower one for three quark flavours). In a more recent calculation with the finite baryonic chemical potential [6], the phase boundary and critical parameters are much better defined, and a critical point appears, as shown in Fig. 2.4. In the region below this point, shown with the dashed line, the transition is second-order, or continuous (*cross-over*). This

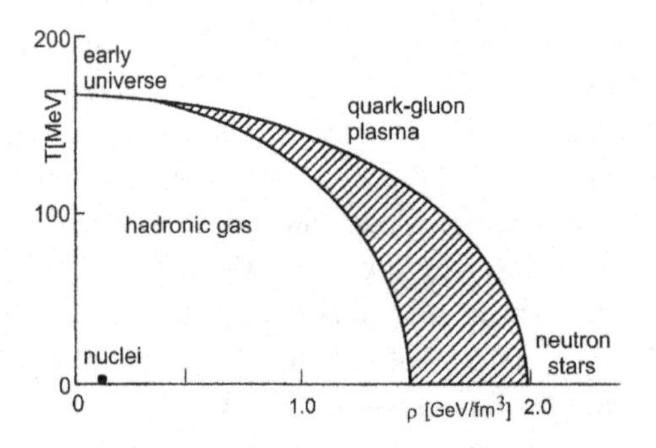

Fig. 2.3 A "classical" phase diagram for strongly interacting matter. Numbers along both axes are approximate.

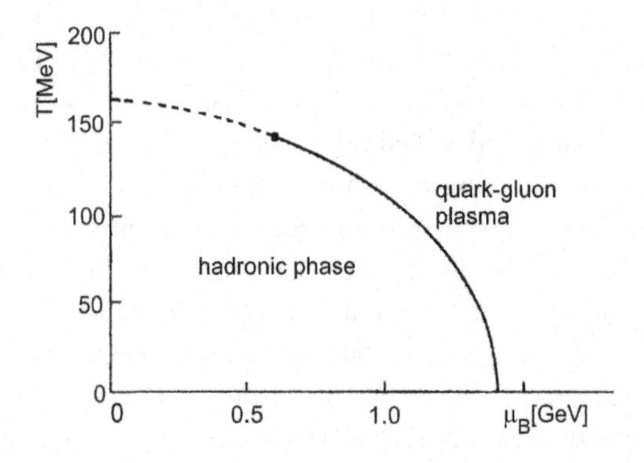

Fig. 2.4 Realistic phase diagram with the critical point.

situation is, unfortunately, not favourable for experimental investigations as in such type of transition no sharp change in the system parameters, indicative of a phase transition, can be expected.

Going further into details, the results of the LQCD calculations relevant to the QGP search concern the order of the phase transition, the value of the critical temperature, other thermodynamic parameters such as pressure or energy density, the screening length in the plasma phase, and thermal effects on hadronic masses, widths and decay constants.

In the pure gluon theory the phase transition is first order. It remains first order if massless quarks of two flavours (u, d) or three flavours (u, d, s) are included. However, with two massless quarks (u, d) and finite mass m_s of the strange quark the transition becomes second order, tending to continuous transitions (*cross-over*) for high s-quark mass. (The second order transition is obtained for m_s of about half the physical mass). This is illustrated in Fig. 2.5 which displays regions of various order of the transition as function of quark masses.

Figure 2.6 shows the phase diagram containing experimental points from SIS, AGS, SPS and RHIC.

Figure 2.7 shows the variation of energy density with increasing temperature, showing a step-like behaviour at T_c, characteristic for a phase

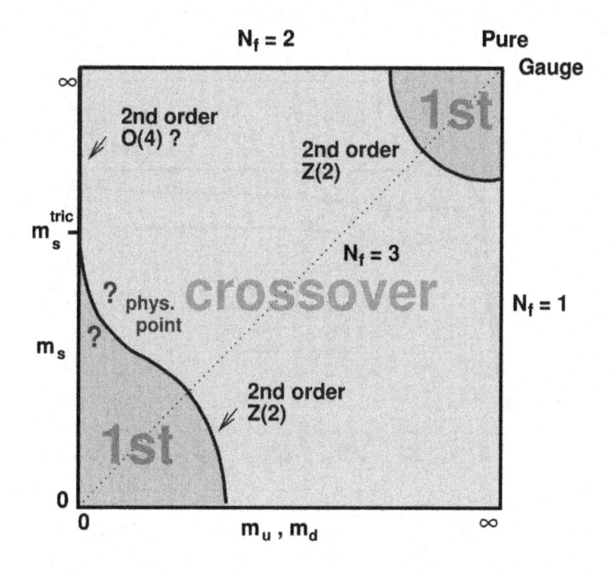

Fig. 2.5 Order of transition as function of quark masses (LQCD results).

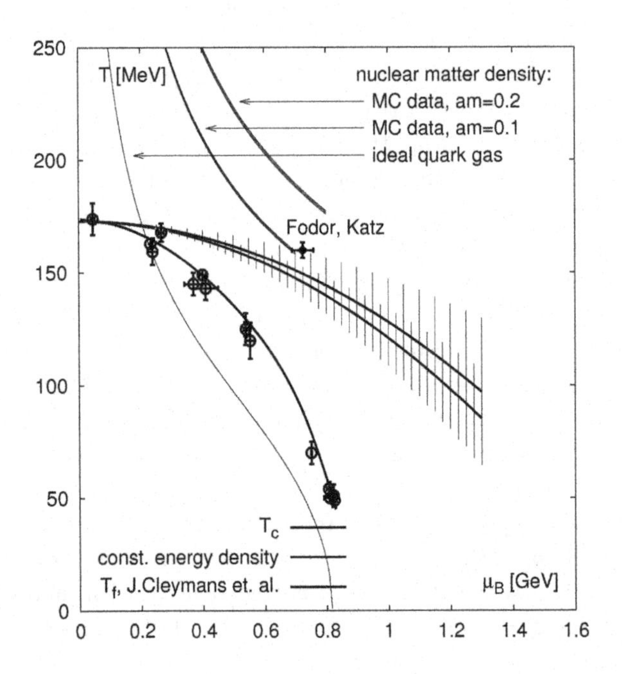

Fig. 2.6 Phase diagram with experimental points (from Ref. [7]).

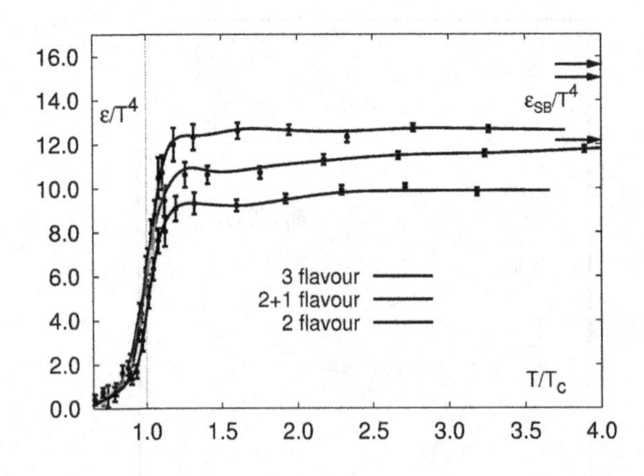

Fig. 2.7 Energy density as function of temperature (LQCD calculations, from Ref. [7]).

transition. This reflects a sudden increase of the number of degrees of freedom when a transition from a hadron gas (HG) to the quark-gluon plasma (QGP) occurs: for a hadron gas this number is between 3 for a pure pion gas, and 10–15 if resonances are included, while for a two-flavour quark-gluon plasma it reaches 37. A straightforward calculation of the number of degrees of freedom for a two-flavour QGP gives

$$g = 2 \times 2 \times 2 \times 3 + 8 \times 2 = 24 + 16 \tag{2.12}$$

Here the first term describes quarks (consecutive weights of two take care of the existence of quark and antiquark states, two flavours, and two spin states, and the weight of three is for colours), and the second term describes gluons which come in eight colours with two possible spin orientations. The contribution of quarks in this formula should, however, be multiplied by the factor $\frac{7}{8}$ in order to account for a difference in the normalization of Boltzmann and Fermi distribution functions, what finally gives

$$g = 21 + 16 = 37 \tag{2.13}$$

An analogous calculation for a three-flavour QGP yields $g = 47.5$.

Let us note that the values of energy densities displayed in Fig. 2.7 do not reach the Stefan-Boltzmann limit $\varepsilon_{\rm SB}$ which corresponds to the ideal gas. This can point out towards some residual interactions in the real system. In fact, recent results from RHIC indicate that the new phase of matter behaves rather like an (almost ideal) liquid than like a gas [8]. Some new ideas about quark-gluon plasma will be discussed in Chap. 21.

References

[1] K. T. Brinkmann, P. Gianotti and I. Lehmann, *Nucl. Phys. News* **16** (2006) 15.
[2] J. C. Collins and M. J. Perry, *Phys. Rev. Lett.* **34** (1975) 1353.
[3] N. Cabibbo and G. Parisi, *Phys. Lett. B* **59** (1975) 67.
[4] E. V. Shuryak, *Phys. Lett. B* **78** (1978) 150.
[5] W. M. Yao *et al.* (Particle Data Group), *J. Phys. G: Nucl. Part. Phys.* **33** (2006) 1.
[6] Z. Fodor and S. D. Katz, *JHEP* **0404** (2004) 050.
[7] F. Karsch, *Nucl. Phys. A* **698** (2002) 199c.
[8] *Quark-Gluon Plasma — New Discoveries at RHIC*, *Nucl. Phys. A* **750**(1) (2005).

Chapter 3

Basic Properties of Atomic Nuclei

3.1 Static properties

We will discuss here only nuclear properties which are of importance in relativistic nuclear physics. In experiments discussed in this book the projectile energies exceed the binding energy of nucleons in nuclei, or excited level spacing in nuclei, by a few orders of magnitude, and thus the details of the internal nuclear structure are of no importance for the studied reactions. What remains of importance is only the nucleon composition of nuclei, and their spatial density distribution, or, in simpler terms, their size and shape. Our discussion, and examples, will be limited to stable or very-long-lived nuclei which can be used as nuclear projectiles or nuclear targets.

A nucleus ${}^{A}_{Z}X^{N}$ characterized by the atomic number Z and the mass number A is composed of Z protons and $N = A - Z$ neutrons. For many light nuclei $N = Z$, or $Z/A = 0.5$ (the heaviest such stable nucleus is ${}^{40}_{20}$Ca), for heavier nuclei $N > Z$, or $Z/A < 0.5$ (e.g. for ${}^{208}_{82}$Pb $N = 1.54Z$ and $Z/A = 0.39$). The Z/A ratio determines the energy to which a given nucleus can be accelerated in a given machine, as the accelerating electrical field acts only on the charge. For example, the Super Proton Synchrotron (SPS) at CERN accelerates protons to $E = 400$ GeV, nuclei with $Z/A = 0.5$ to 200 GeV/nucleon, and ${}^{208}_{82}$Pb nuclei only to 158 GeV/nucleon.

The spatial density distribution of nuclei can be obtained from scattering experiments. Electrons are the best probes as they are point-like objects, and a systematic study of high energy electron scattering off various nuclei was conducted in the late 1950-ies in Stanford, USA, by the Hofstadter group (Nobel Prize 1961). As electrons probe the electric charge, this method allows to obtain the spatial distribution of protons. It was found that this distribution has the shape shown in Fig. 3.1 and can be

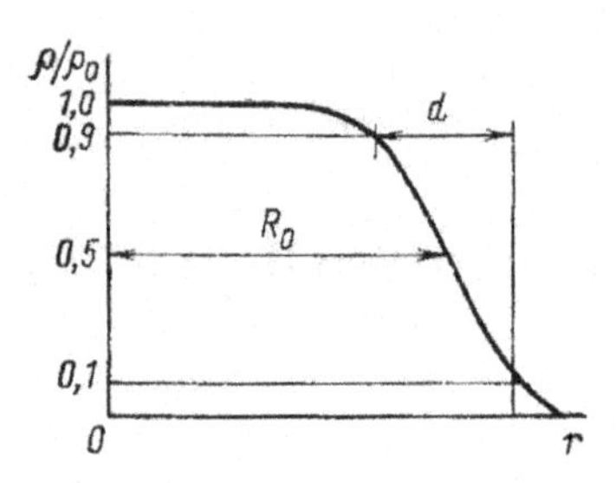

Fig. 3.1 The Saxon–Woods density distribution.

well approximated [1] by the Saxon–Woods (or Fermi) formula

$$\rho(r) = \rho_0/(1 + e^{(r-R)/z}) \tag{3.1}$$

Here ρ_0 is the density in the central plateau region, R is the mean electromagnetic radius of the nucleus (radius at $\rho = 0.5\ \rho_0$), and z characterizes the thickness of the surface layer. The latter is usually replaced by the quantity d, defined as the distance between the points at $0.1\rho_0$ and $0.9\rho_0$. In terms of the parameter z it is equal to $d = 4z \ln 3$. The experimental results indicate that the mean electromagnetic radius of nuclei increases with their mass number as

$$R = (1.07 \pm 0.02)A^{1/3}\ \text{fm} \tag{3.2}$$

where 1 fm $= 10^{-15}$ m , and the thickness of the surface layer, sometimes called *diffuseness*, is approximately the same for all nuclei

$$d = (2.4 \pm 0.3)\ \text{fm} \tag{3.3}$$

In light nuclei, up to about carbon, the central plateau of density disappears, and the density distribution resembles a Gaussian. Scattering experiments using hadrons (protons, neutrons) instead of electrons lead to somewhat higher values of nuclear radii

$$R \cong 1.21A^{1/3}\ \text{fm} \tag{3.4}$$

The nuclear matter density distribution is usually obtained from that for protons by multiplying it by the ratio A/Z. Figure 3.2 shows density distributions $\rho(r)$ for various nuclei. One can see that, except for the lightest nuclei, the density of nucleons in the central region of nuclei is constant and equal to

$$\rho_0 = 0.17\ \text{nucleons/fm}^3 \tag{3.5}$$

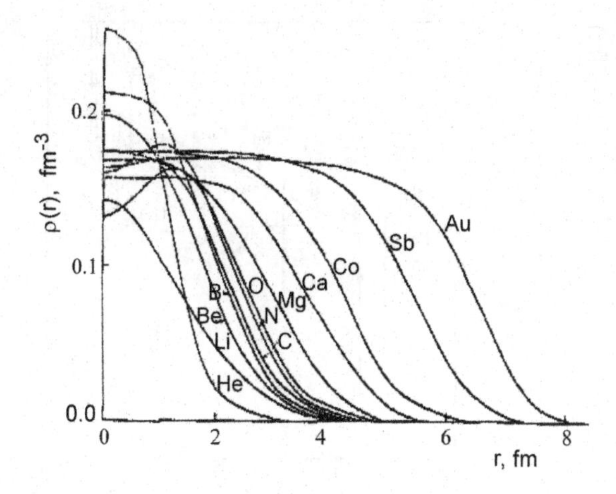

Fig. 3.2 Density distribution functions for various nuclei.

This value is often referred to as *density of nuclear matter*. The above procedure assumes, however, that spatial distributions of protons and neutrons are the same, which is only approximately true. There is evidence from studies of X-ray spectra of antiprotonic atoms, and also from radiochemical experiments, that for heavy nuclei the neutron distribution extends further outwards. These results seem to be compatible with the density profile of the "halo" type, where half-radii for neutron and proton distributions are equal, and distributions differ only by their diffuseness [2]. Larger diffuseness of the neutron distribution leads to an excess of neutrons over protons on the periphery of the nucleus and causes a small difference in the r.m.s. radii for neutrons nad protons: $r_n^{\rm rms} > r_p^{\rm rms}$. The difference in r.m.s. radii, Δr_{np}, was recently found to increase linearly with the relative neutron excess $(N - Z)/A$, reaching 0.2 fm for the heaviest nuclei studied — Fig. 3.3. There are indications that for light nuclei, such as carbon, the opposite situation might occur, i.e. the spatial distribution of neutrons is narrower than that of protons, this due to repelling Coulomb forces acting between the latter.

The Saxon–Woods formula Eq. (3.1) assumes a spherical density distribution. Not all nuclei are spherical, however. A non-spherical charge distribution generates a non-zero electric quadrupole moment[a] Q which

[a]Higher order moments have also been detected in some nuclei, the related deviations from a spherical shape are, however, very small, and will not be discussed here.

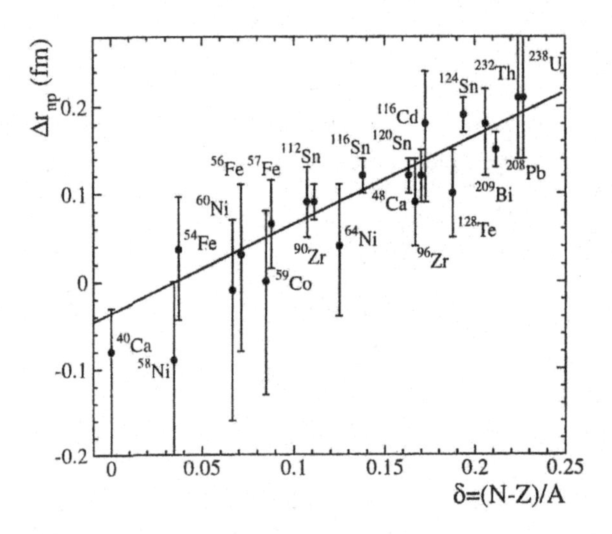

Fig. 3.3 Difference of mean radii of neutron and proton distributions in various nuclei as function of the neutron excess (from Ref. [3]).

can be measured in electron scattering experiments. Assuming the simplest shape of an ellipsoid of rotation with half-axes $R_{x.y}$ and R_z, the quadrupole moment can be expressed as [1]

$$Q = \frac{2}{5}eZ(R_z^2 - R_{x,y}^2) = \frac{6}{5}eZR^2(\Delta R/R) \tag{3.6}$$

where $R = (R_z + 2R_{x,y})/3$ is the average nuclear radius, and $\Delta R = R_z - R_{x,y}$. A negative value of Q means an oblate, or flattened, shape, a positive value of Q means a prolate, or elongated, shape — Fig. 3.4. The

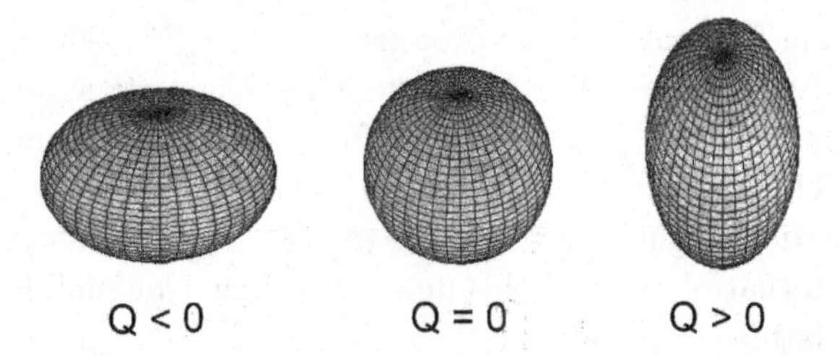

Fig. 3.4 Nuclear shapes and the corresponding electric quadrupole moments.

relative deviation from the spherical shape, $\Delta R/R$, is usually of the order of a few percent only, but for some nuclei it reaches $(25 - 30)\%$. For most nuclei the quadrupole moments have positive values, indicative of an ellipsoid elongated along the z-axis. Figure 3.5 shows the nuclear quadrupole moment systematics: the minima close to the "magic" numbers of nucleons (Z or N = 2, 8, 20, 28, 50, 82, 126) can be clearly seen. Speaking more

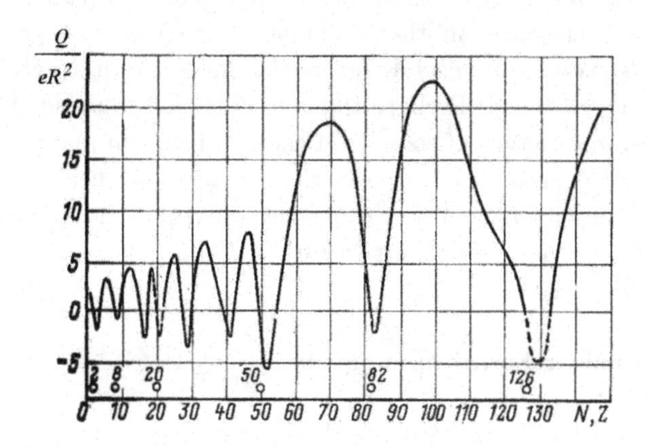

Fig. 3.5 Nuclear electric quadrupole moments systematics (from Ref. [4]).

precisely: just below a magic number $Q < 0$ and small, at the magic number $Q = 0$, and above it $Q > 0$ and rises. Highly elongated nuclei can be found among the lantanides (A = 150–190), and actinides ($A > 220$).

In an unpolarized target nuclei are randomly oriented in space, and thus a projectile impinging upon a target made of non-spherical nuclei encounters variable nuclear matter thickness, what would result in fluctuations in various reaction characteristics, e.g. in the number of produced particles. Uranium-238 ($^{238}_{92}U$) is an example of a strongly deformed nucleus. Its shape can be described as $R = R_o(0.91 + 0.27\cos^2\theta)$, what means that $R_z/R_{x,y} = 1.29$. Such a large deviation from spherical shape should considerably influence the reaction processes. A relatively large broadening of the distribution of the transverse energy flux, $d\sigma/dE_T$, of secondary particles from collisions of 200A GeV sulphur projectiles with uranium target as compared to lead target[b] was indeed observed in a measurement performed at CERN by the NA34/HELIOS Collaboration [5]. For ^{32}S+^{238}U collisions, events in the extreme tail of the E_T distribution are those in which the target nucleus is aligned with the beam. An estimate shows that a longitudinally aligned uranium-238 nucleus produces an E_T correspond-

[b] Lead-208 is a spherical nucleus as a "doubly-magic" one.

ing to a spherical nucleus with atomic number $A \approx 400$. In collisions of two ^{238}U nuclei still larger fluctuations should be expected as in a tip-to-tip collision with impact parameter $b = 0$ the effective thickness of interacting nuclear matter will be almost 50% higher than for a random orientation. This would be an attractive choice for future collider experiments [6].

As it has been stated at the beginning of this chapter, the details of the internal structure of nuclei do not reveal themselves in reactions in the multi-GeV energy range. Perhaps the only deviation from this rule is the internal cluster structure of some light nuclei which can manifest itself in fragmentation processes. As an example one can quote here an enhanced probability for fragmentation of ^{12}C nucleus into α-particles (or ^{4}He nuclei) [7], indicative for the α-cluster structure of ^{12}C.

3.2 The nuclear Fermi momentum

Nucleons in nuclei are not still but find themselves in a chaotic motion called the *Fermi motion.* The relevant description is the Fermi-gas model. It assumes that nucleons fill subsequent energy levels in a potential well. Because of the exclusion principle, each level can be occupied by at most two nucleons of the same kind (protons or neutrons), the energy levels being filled independently for the two nucleon species. The highest energy level is called the *Fermi level*, and corresponds to the nucleon kinetic energy of the order of 20–40 MeV (the lower value is for light nuclei, the higher one is for heavy nuclei). The existence of the Fermi momentum in nuclei has been experimentally verified in quasi-elastic electron scattering experiments [8]. The energy distribution of electrons scattered off nuclei show a broad peak which is a consequence of the internal motion of nucleons in nuclei. From the width of the peak the value of the nucleon Fermi momentum, p_F, can be obtained. Fermi momenta for various nuclei are listed in Table 3.1. The uncertainty of the quoted p_F values is ± 5 MeV. The Fermi momentum is roughly constant for nuclei heavier than Ni, and close to the corresponding value in infinite nuclear matter with density $\rho = 0.17$ fm^{-3}:

$$p_F = (3\pi^2 \rho/2)^{1/3} = 270 \text{ MeV/c or } 1.37 \text{ fm}^{-1} \tag{3.7}$$

The distribution of the nucleon internal momenta in nuclei is usually assumed to be a step function $f(p) = c\,\theta(p_F - p)$, however for light nuclei use of a modified Gaussian has been also proposed [9].

Table 3.1 Values of Fermi momentum for various nuclei (from Ref. [8]).

Nucleus	p_F,MeV/c
^{6}Li	169
^{12}C	221
^{24}Mg	235
^{40}Ca	251
^{nat}Ni	260
^{89}Y	254
^{nat}Sn	260
^{181}Ta	265
^{208}Pb	265

The Fermi momentum influences nuclear reactions, also those occuring at high energies. In a collision of an elementary projectile with a nucleon at rest, the c.m.s. energy is uniquely defined (see Sec. 4 of Appendix A), and so is the energy available for the production of new particles. In a collision with a nucleon in a nucleus, however, the Fermi motion smears the effective energy, what, in particular, lowers the threshold energy for the production processes. As an example we can recall the famous experiment at the Bevatron in which antiprotons were discovered [10] (Nobel Prize 1959). The minimum beam energy for the production of a proton-antiproton pair in a proton-proton collision is 5.6 GeV. However, with a copper target used in this experiment, one can estimate that due to the Fermi momentum the effective threshold energy could be lowered to about 4.3 GeV. Indeed, antiprotons were detected already at beam energies below 5 GeV.

Figure 3.6 shows the distribution of the true nucleon-nucleon centre-of-mass energies for nucleons bound in ^{12}C nuclei colliding with the nominal c.m.s. energy of 20A GeV for two different assumptions about the shape of the intranuclear momentum distribution: the Fermi gas model and the Gaussian parameterization.

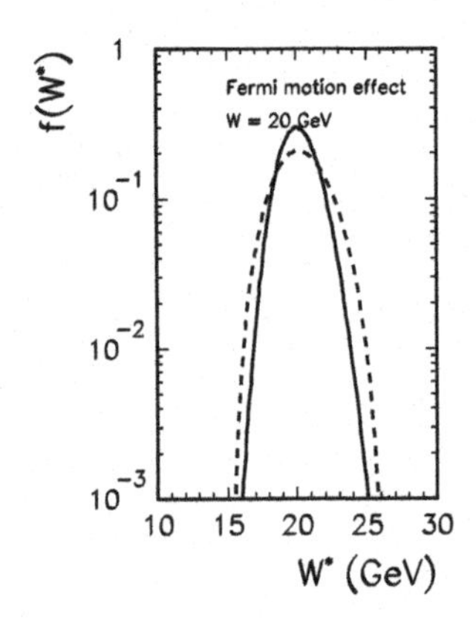

Fig. 3.6 Distribution of the true nucleon-nucleon centre-of-mass energies in a collision of two carbon nuclei for the Fermi gas model (dashed) and the Gaussian parameterization (solid) for nominal energy $W = 20$ GeV (from Ref. [11]).

One can see that widening of the energy distribution is substantial — up to about 20%. A similarly large effect should persist at higher energies, also at LHC energies, what is often being forgotten.

References

[1] H. Enge, *Introduction to Nuclear Physics* (Addison-Wesley, 1972).
[2] P. Pawłowski and A. Szczurek, *Phys. Rev. C* **70** (2004) 044908.
[3] A. Trzcińska *et al.*, *Phys. Rev. Lett.* **87** (2001) 082501.
[4] K. N. Mukhin, *Eksperimentalnaya Yadernaya Fizika*, 4th edn., Energoatomizdat (1983).
[5] C. Akesson *et al.* (HELIOS Collaboration), *Phys. Lett. B* **214** (1988) 295.
[6] C. Nepali, G. Fai and G. Keane, *Phys. Rev. C* **73** (2006) 034911.
[7] A. Korejwo *et al.*, *J. Phys. G: Nucl. Part. Phys.* **26** (2000) 1171; *ibid.* **28** (2002) 1199.
[8] S. Frullani and J. Mougey, *Adv. Nucl. Phys.* **14** (1984) 1.
[9] W. Cassing *et al.*, *Z. Phys. A* **349** (1994) 77.
[10] O. Chamberlain *et al.*, *Phys. Rev.* **100** (1955) 947.
[11] A. Szczurek and A. Budzanowski, *Mod. Phys. Lett. A* **19** (2004) 1669.

Chapter 4

Sources of Relativistic and Ultrarelativistic Nuclei

4.1 Cosmic rays

Before the advent of high energy heavy ion accelerators, cosmic rays constituted the only source of very energetic nuclei. Near to the Earth, but above the atmosphere, cosmic rays consist of 86% protons, 13% helium nuclei, and about 1% heavier nuclei. The latter are mainly light nuclei (up to Si), and Fe nuclei. The energy spectrum of some of these components is shown in Fig. 4.1. As it can be seen from this Figure, above a few GeV/nucleon the cosmic ray energy spectrum falls down steeply. It approximately follows the power law $dI/dE \propto E^{-2.7}$. This steep fall means that at very high energies the intensity is so low that only single events can be recorded. For example, at 10^{15}eV the cosmic ray flux is only 1 particle/m^2 year. Above this energy the exponent is even larger. Nevertheless, particles with energies up to about 10^{20}eV have been detected by recording the so-called *extensive air showers* (EAS) in devoted large-area experiments.

Due to interactions in the atmosphere, the elemental composition of cosmic rays changes with the penetration depth, the relative amount of heavier species decreases (see Fig. 4.2), and at the sea level they almost completely disappear. The energy of cosmic ray particles is also being degraded, this being the reason to perform cosmic ray experiments in the stratosphere (balloons, artificial satellites), or at high mountain tops. Apart from a low flux of high energy cosmic ray nuclei, also their energy cannot be precisely determined, what affects the reliability of event analysis. Nevertheless, some basic features of nuclear interactions at very high energies (interaction cross sections, limited transverse momentum of produced particles) have been already established in cosmic ray experiments in the 1950s.

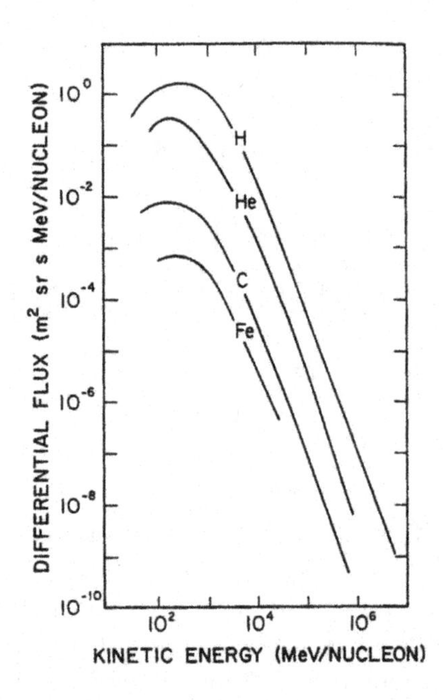

Fig. 4.1 Differential energy spectrum of some of the major components of cosmic rays (from Ref. [1]).

4.2 Accelerators

Experiments with relativistic heavy ions under fully controlled conditions can be made at accelerators. In the late 1960s it was realized that acceleration of heavy ions in the old few-GeV proton synchrotrons would prolong their active life and open a new field of research and applications. In 1970 deuterons and alphas were successfully accelerated in the Dubna synchrophasotron to the energy of about 5 GeV/nucleon, or, using the present-day notation, to $5A$ GeV. Within the next year various light nuclei have been accelerated in the Bevatron in Berkeley and in Saturne in Saclay to energies between $1A$ and $2A$ GeV. Heavier ions with the same ratio $Z/A = 0.5$ (i.e. nuclei up to ${}^{40}_{20}\mathrm{Ca}$) could be accelerated in the same regime as deuterons and alphas — in fact ${}^{14}_{7}\mathrm{N}^{7+}$ ions were very soon accelerated in Berkeley to $2.1A$ GeV.

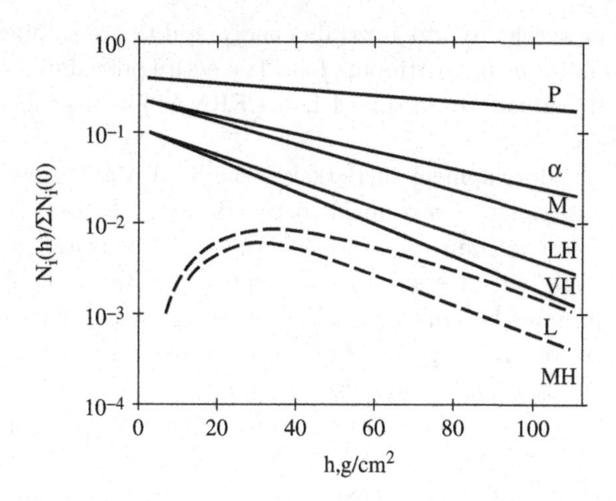

Fig. 4.2 Fluxes of various cosmic ray nuclei in the atmosphere as a function of penetration depth. From top to bottom: protons (p), helium (α), Z = 6–9 (M), Z = 10–14 (LH), Z = 20–28 (VH), Z = 3–5 (L), and Z = 15–19 (MH). The last two groups of nuclei, shown with dashed lines, are not present in the primary cosmic radiation, and result from secondary processes in the atmosphere (from Ref. [2]).

Technical problems arising when heavy ions are being fed to a proton synchrotron are threefold:

(i) ion beams require a lower accelerating radiofrequency which may fall outside the tuning range of the r.f. generator — this problem can be circumvented by accelerating initially on a higher harmonic;
(ii) the beam intensity must be high enough as to match the beam control and monitoring system of the accelerator;
(iii) the vacuum in the accelerator ring must be good enough not to cause excessive losses in the beam intensity as multicharged ions, especially if not fully stripped of electrons, show very high recombination cross sections. The problem of obtaining high vacuum is most difficult in weak-focusing machines with large-volume vacuum chambers.

Obviously, a good ion source is the essential element. Electrons from inner atomic shells are strongly bound and an ion source should possess a high ionization power in order to remove them. In the widely used Electron Cyclotron Resonance (ECR) sources it is an intense electron beam circulating in a strong solenoidal magnetic field which constitutes the ionizing medium. Nevertheless, for really heavy ions, such as Au or Pb, only partial ionization can be achieved in such a source. Thus partly stripped ions should be accelerated to a certain energy, and the remaining electrons are removed later, using a stripping foil. The entire procedure of obtaining fully stripped lead ions, or lead nuclei, at CERN will be described below as an example.

In a further development in Berkeley, the $8.5A$ MeV heavy ion linear accelerator Superhilac was coupled to the Bevatron as its injector, thus forming the Bevalac complex. Cryogenic linings were installed inside the Bevatron vacuum chamber in order to improve the vacuum. With these, ions up to uranium became available to experimenters. The intensity of $^{238}_{92}U^{68+}$ beam was about 10^6 ions per pulse at energy of $1.1A$ GeV, with the repetition rate of about 10 pulses per minute.

In Dubna, cryogenic panels were installed inside the vacuum chamber of the Synchrophasotron in order to improve the vacuum. Acceleration of ions up to krypton ($Z = 28$) was achieved.

In Saclay the entire synchrotron ring magnet was replaced with a new small-aperture strong-focusing one, called Saturne-II, what allowed acceleration of heavier ions.

All three old synchrotrons: the Bevalac, Synchrophasotron and Saturne are no longer in operation. In the energy range of a few GeV/nucleon there are now two machines: Nuclotron at JINR in Dubna and SIS at GSI in Darmstadt.

Nuclotron is a strong-focusing cryogenic synchrotron, of 251 m circumference, built in a tunnel below the old synchrophasotron magnet, and commissioned in 1993. It accelerates various ions up to krypton to energies up to about $4.2A$ GeV. Initially it operated with internal target, the beam extraction was realized in the year 2002. The goal is to achieve the energy of $6A$ GeV, and to handle heavier ions (up to uranium). This should be realized by an upgrade comprising the installation of a more efficient ion source and addition of an intermediate booster synchrotron.

SIS is also a strong-focusing synchrotron, of 216 m circumference, commissioned in 1990, and using the GSI heavy ion linear accelerator UNILAC as injector. SIS provides beams of almost any ions, with energies up to $2A$ GeV for light ions and $1A$ GeV for uranium, and with relatively high intensities of 10^{11} for light ions, 10^{10} for krypton, and 10^9 for uranium. Fast or slow beam extraction are possible.[a]

---

[a]SIS is considered as the basic element of a new project developed at GSI: the Facility for Antiproton and Ion Research (FAIR). This will be a collider with the double ring of 1100 m perimeter. It should provide beams of all ions up to uranium with unprecedented intensity (10^{12} per cycle) and quality. With colliding beams of up to $20A$ GeV each,

In autumn 1986 relativistic heavy ion programmes started at BNL and CERN. In Brookhaven, the existing tandem Van-de-Graaff accelerator was connected by a 600 m long transfer line to the Alternating Gradient Synchrotron (AGS). On October 20, 1986, ^{16}O ions were accelerated to 14.6A GeV, with the intensity of 4×10^8 ions per pulse, and ^{28}Si ions were accelerated to the same per-nucleon energy in 1987. A booster synchrotron, built in the next years, allowed to obtain heavier beams, up to Au.

At CERN, the ECR ion source commissioned in Grenoble was coupled to the old Linac-1 via a special radiofrequency quadrupole unit (RFQ) built at LBL, and a beam of oxygen ions was fed into the Super Proton Synchrotron (SPS). On September 7, 1986, an ^{16}O beam reached the energy of 200A GeV, and on September 25, 1987, ^{32}S ions were accelerated to the same per-nucleon energy. The acceleration process lasts about 10 seconds and involves five consecutive accelerators: RFQ Linac, Linac-1, Booster Synchrotron, PS and SPS. The CERN accelerator complex is shown in Fig. 4.3. The new ion source allowed to obtain a beam of ^{208}Pb ions with

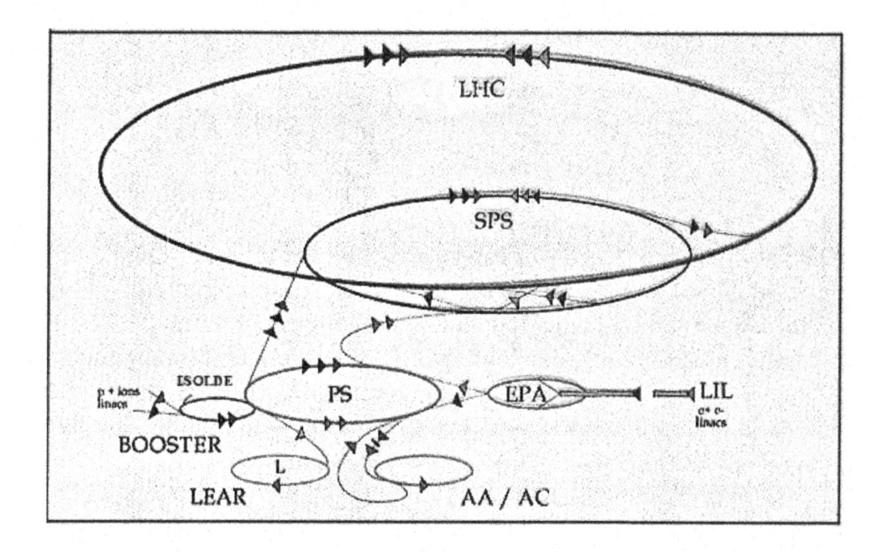

Fig. 4.3 Scheme of the CERN accelerator complex, including the LHC. The drawing is not exactly to scale, the circumference of SPS being 6.2 km, and that of LHC is 27 km.

---

the CERN SPS energy range will be covered and surpassed, at the same time allowing for high precision measurements and studies of very rare processes. It would take eight years to build the FAIR facility.

energy of 158A GeV in 1994. Recently, indium ($Z = 49$) ions have been accelerated in order to study interactions of medium-size nuclei. It might be of interest to describe in more detail the procedure of obtaining fully stripped Pb ions. This is shown in the diagram below. Let us note that after the first stripping $Pb^{28+} \to Pb^{53+}$ at 4.2A MeV, the second (and final) one $Pb^{53+} \to Pb^{82+}$ is performed at the energy as high as 4.25A GeV.

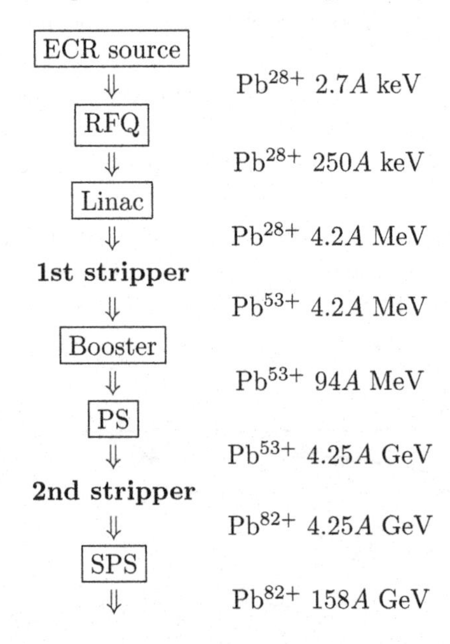

Further developments in investigating collisions of heavy ions at still higher energies are connected with ion colliders. It should be recalled that light ions: deuterons and alphas, were accelerated to 15.7A GeV in the CERN Intersecting Storage Rings (ISR). With luminosity in excess of 10^{28} cm^{-2}s^{-1}, several experiments were performed in the years 1980–1983. Running the ISR with heavier ions would certainly have been feasible, but this excellent machine was closed in 1984 for economic reasons.

In the year 2000 the Relativistic Heavy Ion Collider (RHIC) was commissioned in Brookhaven. It uses the AGS as the injector. RHIC collides Au beams of 100A GeV each, what means the centre-of-mass energy of 200 GeV per nucleon pair, or ten times higher than that reached with a stationary target at the CERN SPS. The luminosity for Au+Au collisions is 10^{27} cm^{-2}s^{-1}. RHIC can also accelerate protons and lighter ions (e.g. Cu), and is able to work at lower energies, thus covering the entire energy interval above that of the CERN SPS. Two separately powered rings make

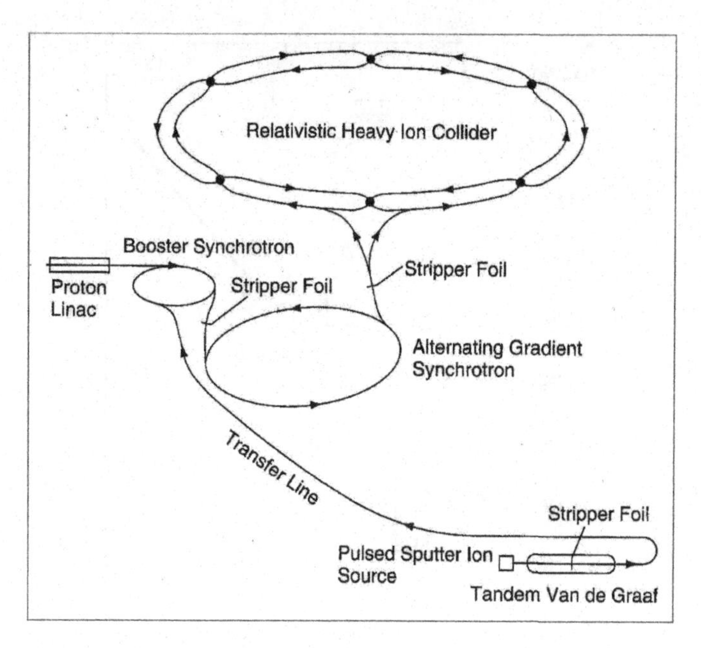

Fig. 4.4 Scheme of the BNL accelerator complex. The circumference of RHIC is 3.8 km. The two rings of the machine cross in six points.

possible the operation with either equal or unequal ion species in the colliding beams, e.g. d+Au. The BNL accelerator complex is shown in Fig. 4.4. An upgrade of RHIC is being considered, aimed at the luminosity increase by a factor of ten by adding the electron beam cooling, and at colliding uranium nuclei. This project has been named RHIC-II.

A new collider, the Large Hadron Collider (LHC) is now being commissioned at CERN, in the 27 km long undergound tunnel which earlier housed the LEP machine (see Fig. 4.3). It will use the CERN SPS as the injector. LHC is primarily a proton collider attaining $\sqrt{s} = 14$ TeV, but acceleration of heavy ions (up to Pb) is also foreseen. For ions the centre-of-mass energy per nucleon pair, $\sqrt{s_{NN}}$, will be 7 TeV for light ions, and 5.5 TeV for Pb ions, or almost 30 times higher than that in RHIC. The first proton beams in the LHC are expected in 2008, and the first ion beams in 2009. The luminosity for lead beams should reach $10^{27}\mathrm{cm}^{-2}\mathrm{s}^{-1}$. As this machine uses "two-in-one" magnets, only "symmetric" ion-ion collisions will be feasible.

Both colliders: RHIC and LHC use cryogenic magnets. The dipole magnets for the LHC are designed for the field of 8.4 Tesla. With 1232 14-meter long dipoles and some 300 quadrupoles and correction magnets,

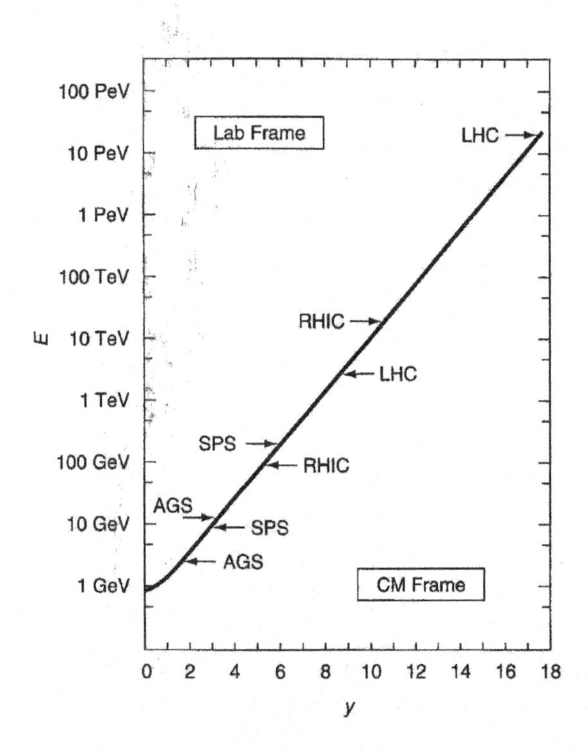

Fig. 4.5 Energies and ranges of rapidity in the laboratory frame and in the centre-of-mass frame for various high energy heavy ion accelerators (from Ref. [3]).

the LHC is not only the biggest accelerator, but also the biggest cryogenic installation ever built.

Figure 4.5 shows a comparison of energies and rapidity ranges for various relativistic heavy ion accelerators, including the LHC. The rapidity of the centre-of-mass frame, $y_{\rm cm}$, is 3.0 for the SPS, 5.4 for RHIC, and will be 8.7 for LHC. One can see that for energies above a few GeV the rapidity range increases as $\ln E$. The total rapidity range in the c.m. frame is $2y_{\rm cm}$, or 6.0, 10.8, and 17.4 units of rapidity correspondingly for the SPS, RHIC, and LHC.

References

[1] R. Silberberg and C. H. Tsao, *Phys. Rep.* **191** (1990) 351.
[2] Z. Włodarczyk, *Proc. 5th Int. Symp. Very High Energy Cosmic Ray Interactions*, Łódź, ed. M. Giler (1989), p. 160.
[3] K. Yagi, T. Hatsuda and Y. Miake, *Quark-Gluon Plasma* (Cambridge University Press, 2005), Chap. 10.

Chapter 5

Detection Techniques

5.1 Fixed-target experiments

In this class of experiments a beam of accelerated nuclei impinges upon a gaseous, liquid, or solid stationary target. Due to Lorentz boost, the secondary particles are collimated in the forward direction, and thus an "almost 4π" acceptance is achieved with installations covering only a relatively small space angle. The typical detector geometry is a conical one, diverging from the target along the incident beam direction.

In early days visual detectors: bubble chambers, nuclear emulsions, or streamer chambers, were in use. Examples of interactions of relativistic heavy ions recorded in such detectors are shown in Figs. 5.1–5.3. These detectors provide a 4π coverage, but the evaluation of the obtained data is very laborious and time-consuming, what results in small statistics of analyzed events — typically hundreds or few thousands.

A main break-through came with the advent of the time projection chamber (TPC) — a large-volume gaseous detector with electronic read-out. With this detector, statistics of millions of recorded events became accessible, what meant a real progress. The NA49 experiment at CERN was among the first to employ TPC's as main detectors, supplemented with some other ones. Figure 5.4 shows the scheme of NA49, and Fig. 5.5 displays a visualization of tracks in the NA49 TPC's. The NA49 set-up was a "multipurpose" detector. It produced a wealth of data on identified particles abundances, longitudinal and transverse spectra and correlations, mainly for Pb+Pb collisions at incident energies between $20A$ and $158A$ GeV, but also for lighter nuclear projectiles, including protons. Another "multipurpose" detector was WA98 (developed from WA80/WA93) which

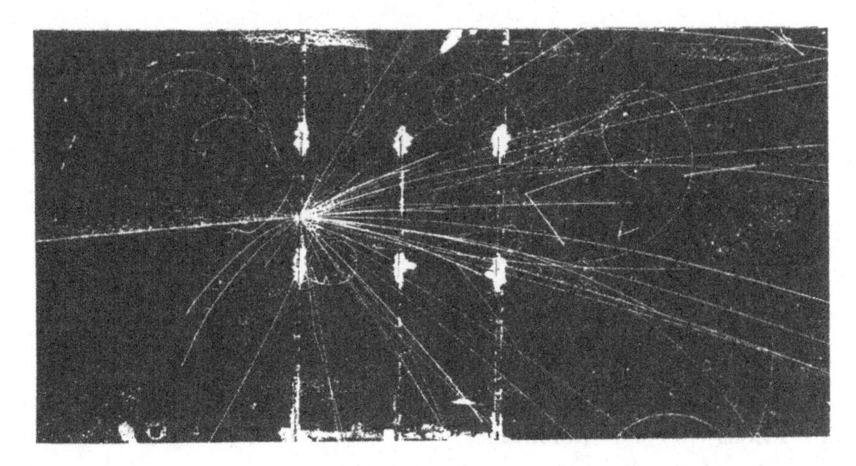

Fig. 5.1 Interaction of a 50 GeV/c (4.2A GeV/c) ^{12}C projectile from the Dubna synchrophasotron in a tantalum target plate mounted inside the propane bubble chamber. The chamber was 2m long, and operated in a magnetic field. This photo covers one-half of the chamber length.

apart of charged particles detected photons and measured their energy using segmented lead glass spectrometer. Its scheme is shown in Fig. 5.6.

Other detectors at the CERN SPS were designed to study more specific aspects of heavy ion reactions. A radial time projection chamber was used in the CERES/NA45 experiment which is dedicated to the measurement of low energy $e^+ e^-$ pairs close to midrapidity, providing full azimuthal acceptance within the pseudorapidity interval $2.1 < \eta < 2.6$.

The WA57/NA57 experiment studied the production of hyperons and antihyperons. It based mainly on silicon detectors (pixel and strip types), operated in the magnetic field of a big dipole magnet.

Special small aperture spectrometers measured single-particle spectra and close momentum correlations of pions, kaons and protons with high resolution (NA44), and studied the mass spectrum of secondary particles (NEWMASS/NA52).

Production of muons was the main objective of the NA50/NA60 experiment. Its main part was the toroidal magnetic spectrometer with a number of multiwire tracking chambers.

In Figs. 5.7–5.11 we show schemes of these detectors, together with their short description. The schemes give evidence for a large variety of technical solutions.

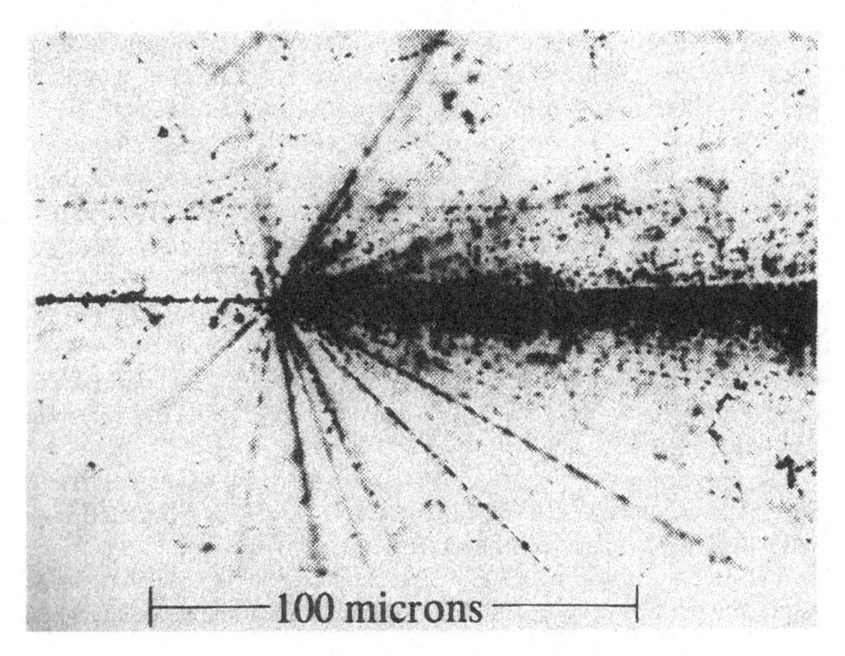

Fig. 5.2 Interaction of a 6.4 TeV (200*A* GeV) ^{32}S projectile from the CERN SPS with a heavy nucleus (Ag or Br) of nuclear emulsion. The scale is shown near te bottom of the picture, giving evidence for the very high resolution of this type of detector.

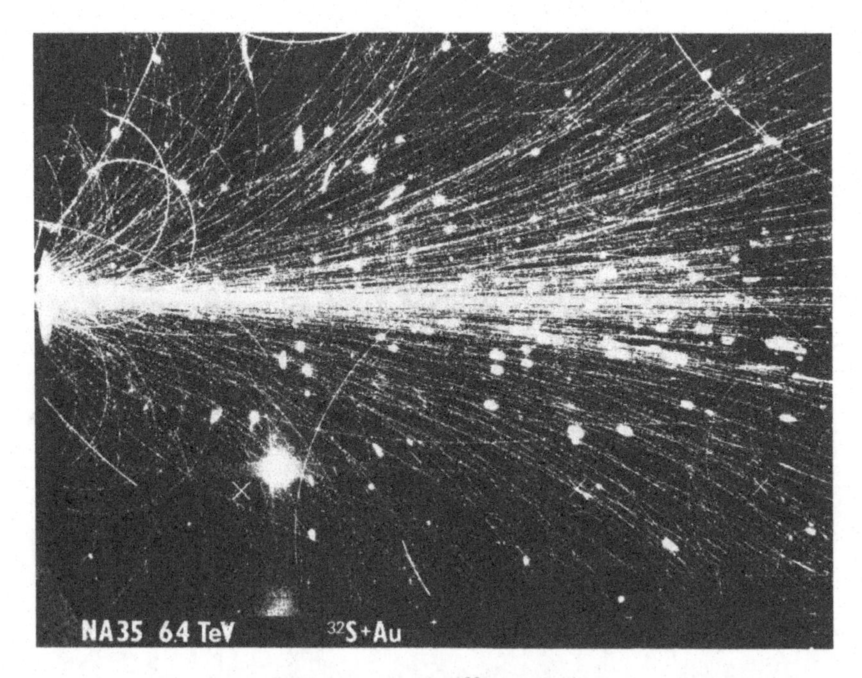

Fig. 5.3 Interaction of a 6.4 TeV (200A GeV) ^{32}S projectile from the CERN SPS in the gold target positioned in front of the entrance window of the 2m long streamer chamber of the NA35 collaboration. The streamer chamber operated in a magnetic field.

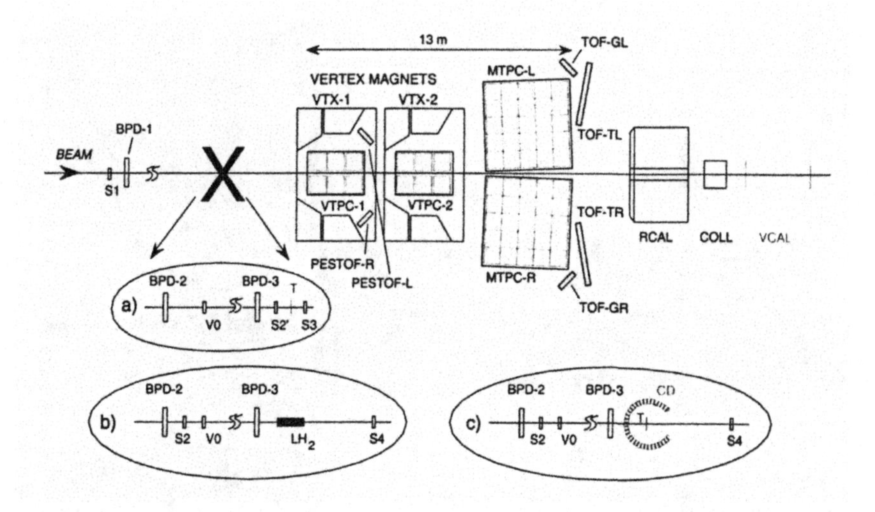

Fig. 5.4 Scheme of the NA49 experiment at the CERN SPS. Two TPC's: VTCP-1 VTPC-2 are placed in magnetic field, and serve as tracking detectors, two bigger TPC's: MTPC-L and MTPC-R, serve mainly for energy loss measurements. Time-of-flight detectors TOF, and the azimuthally and radially segmented calorimeter RCAL complete the set-up. Various configurations of detectors in the vicinity of the target T are shown expanded at the bottom of the Figure. These are specific for (a) A+A, (b) p+p, and (c) p+A collisions.

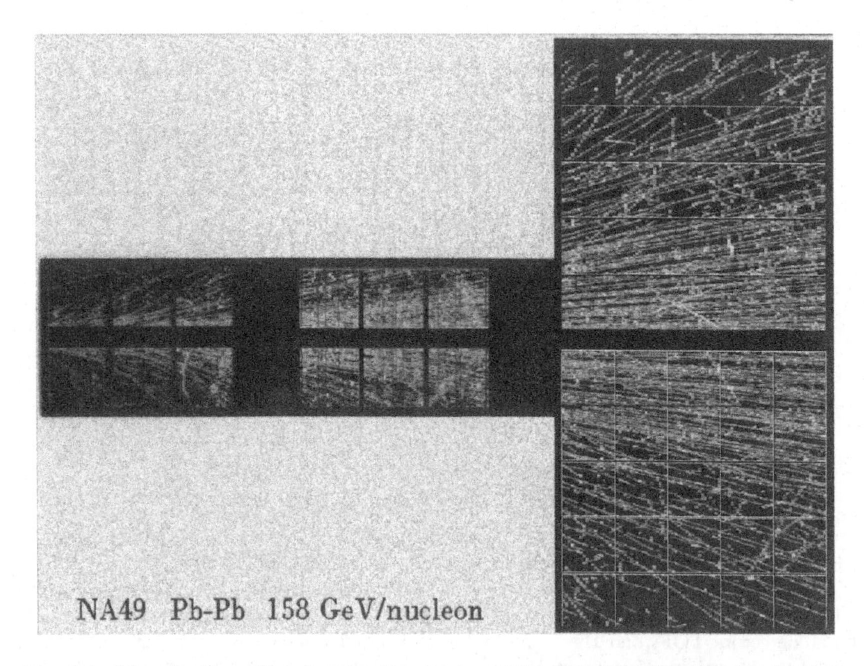

Fig. 5.5 Visualisation of tracks emerging from a 158A GeV Pb+Pb central collision in four time projection chambers of the NA49 experiment at the CERN SPS.

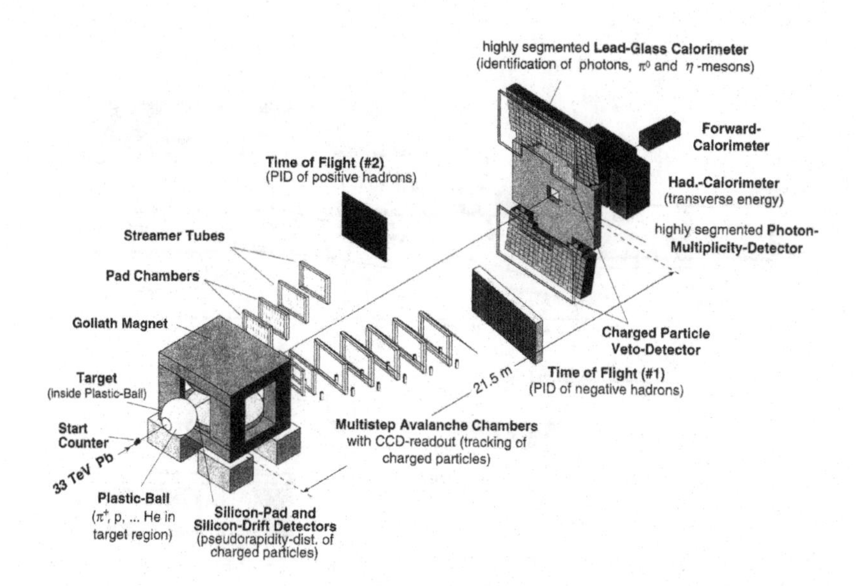

Fig. 5.6 Schematic picture of the WA98 set-up at the CERN SPS. The main components of this set-up are the large acceptance photon and charged hadron spectrometers. Photons are detected, and their energy measured, in the 10,000 modules lead glass array. Negatively charged hadrons are analyzed in the magnetic spectrometer equipped with six multistep avalanche chambers for tracking. Total charged multiplicity is measured with high resolution sislicon detectors, and slow charged particles and nuclear fragments in the target region are identified in the Plastic Ball detector. The latter consists of 655 $\Delta E - E$ modules arranged in a sphere around the target.

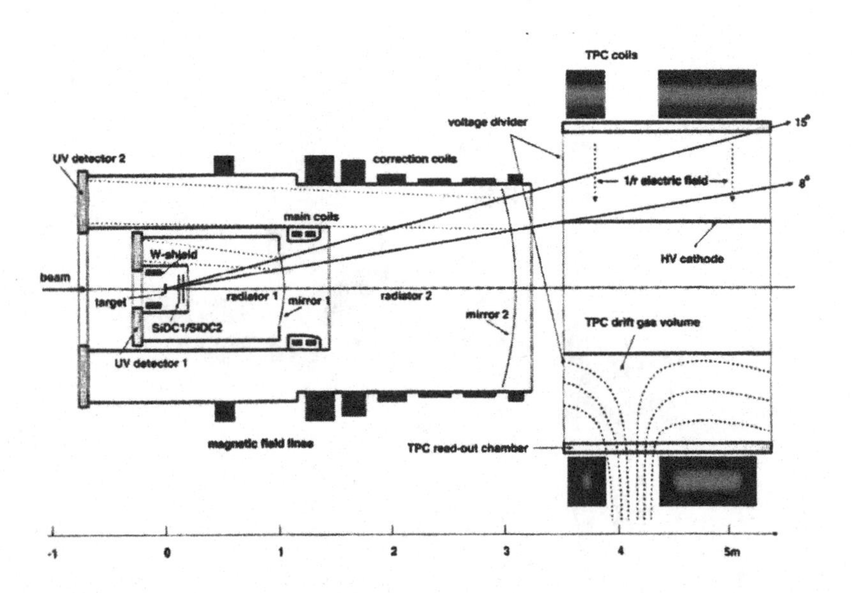

Fig. 5.7 Scheme of the CERES/NA45 experiment at the CERN SPS. The apparatus has a cylindrical symmetry, Two silicon drift detectors SDD provide tracking and vertex reconstruction, two ring imaging Cherenkov detectors RICH give electron identification, and the radial drift time projection chamber TPC, operated inside a magnetic field, measures the momentum and energy loss.

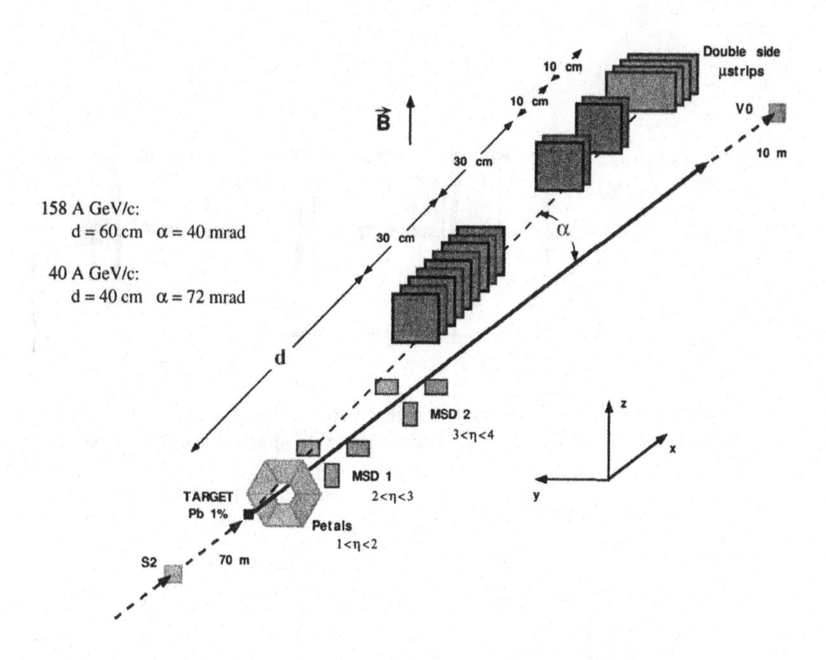

Fig. 5.8 Schematic picture of the NA57 set-up at the CERN SPS. This set-up consists of a telescope made of high granularity silicon pixel detectors and double-sided silicon strip detectors downstream. The entire telescope is placed above the beam line, at a certain angle α, and accepts particles produced near midrapidity, with medium transverse momenta.

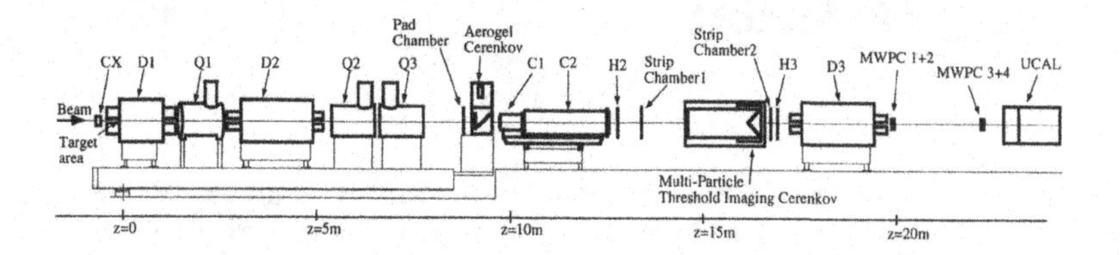

Fig. 5.9 Scheme of the NA44 set-up at the CERN SPS. This is a focusing spectrometer for the measurement of a few particles around midrapidity. The magnet part of the spectrometer uses two dipoles and three quadrupoles. Then particles are tracked and identified with several detectors: multiwire proportional chambers, scintillator hodoscopes, and threshold gas Cherenkov counters.

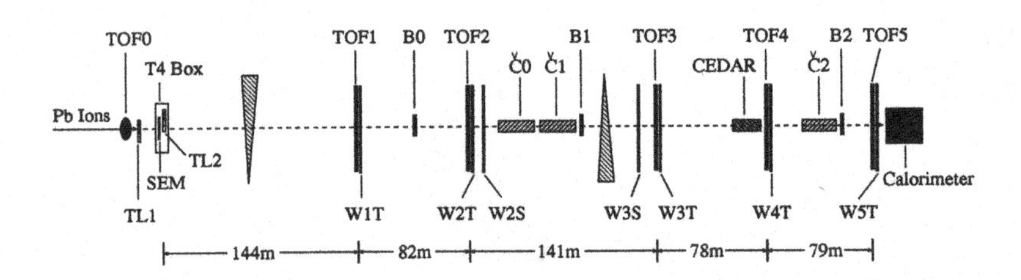

Fig. 5.10 Scheme of the NA52 set-up at the CERN SPS. This experiment uses one of the SPS secondary beam lines as a focusing spectrometer of charged particles close to zero degree production angle. Seven sets of multiwire proportional chambers (W) are used for particle tracking. Five segmented time-of-flight scintillator hodoscopes (TOF), three threshold Cherenkov counters (Č), and one differential Cherenkov detector (CEDAR) are used for particle identification.

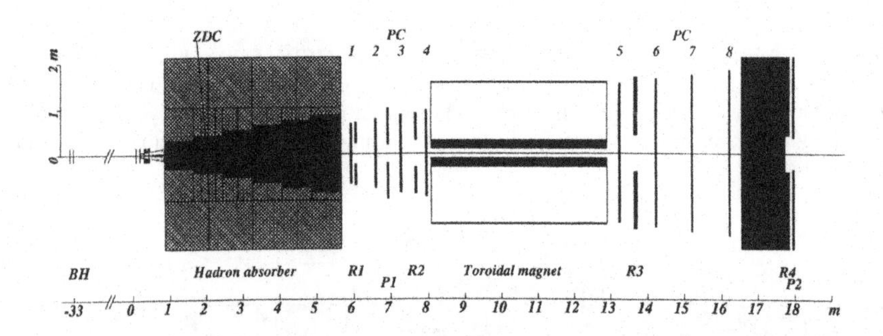

Fig. 5.11 Scheme of the NA50 set-up at the CERN SPS. This detector is composed of the thick hadron absorber and the muon spectrometer with a toroidal magnet and a number of multiwire chambers placed before and after the magnet, serving for muon tracking.

5.2 Experiments at colliders

In colliders two beams accelerated in opposite directions collide in vacuum. The main advantage of colliders is the gain in effective energy of the collision (see Sec. 4 of Appendix A). On the other hand, achieving a high luminosity is difficult (needs a very precise beam collimation and steering), and in order to provide a 4π acceptance, the detector should cover full space angle. The typical detector geometry is a cylindrical one, with the axis of the cylinder aligned with the beam pipe of the accelerator, and the beam collision point lying in the geometrical centre of the cylinder. As the full angular coverage is technically difficult, and also very costly, in many detectors only a fraction of the total space angle is fully instrumented. This is usually an angular interval near 90^0, corresponding to the kinematical region close to midrapidity. Small angle regions, corresponding to large rapidities, are also instrumented, often mainly for triggering purposes. The variety of employed detectors is very wide. In some installations (STAR, ALICE) a cylindrical TPC constitutes the main tracking detector. High resolution silicon detectors are often installed close to the beam pipe of the collider, while transition radiation detectors, ring-imaging Cherenkov detectors (RICH), time-of-flight counters, and various types of calorimeters are installed at larger distances.

In Figs. 5.12–5.15 schemes of the four detectors which operate at RHIC: BRAHMS, PHOBOS, PHENIX, and STAR are shown, together with their short description. Figure 5.16 displays a visualisation of tracks emitted into a transverse slice near midrapidity from an Au+Au collision recorded in the STAR detector.

Finally, in Fig. 5.17 we show the schematic picture of the ALICE detector dedicated to study heavy ion collisions at the LHC, and in Fig. 5.18 the computer simulation of particle tracks emerging from a nuclear collision in ALICE. The ALICE Collaboration gathers about 1,000 physicists and engineers from 100 institutes and laboratories in 30 countries. The total cost of the detector is about 100 million Euros. At the time of writing the ALICE detector is almost completely assembled, and its various components are being tested with cosmic rays. It should be ready to take some reference p+p data at the LHC start-up in automn of 2008, beams of lead ions are expected a year later.

Two other LHC experiments: CMS (**C**ompact **M**uon **S**olenoid) and ATLAS (**A** **T**oroidal **L**HC **A**pparatu**S**), originally dedicated to study p+p collisions, also envisage taking data during the lead ions run. Their main objectives are high p_T spectra and jets, but also heavy flavours, quarkonia (including *upsilon*), and direct photons.

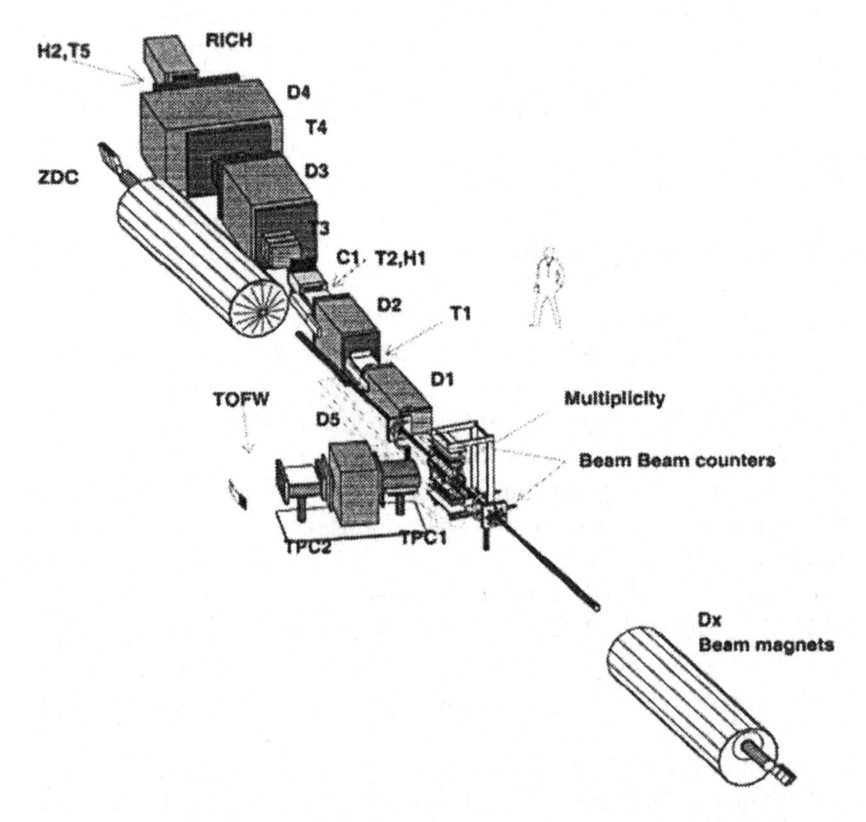

Fig. 5.12 Schematic picture of the BRAHMS detector (BRAHMS = **B**road **RA**nge **H**adron **M**agnetic **S**pectrometer) at RHIC. This set-up is composed of two small aperture spectrometers: one at forward angles, and another one at midrapidity. The forward spectrometer consists of four dipole magnets D1–D4, two time projection chambers T1 and T2, three drift chamber modules T3–T5, time of flight hodoscopes H1 and H2, a threshold Cherenkov counter C1, and a ring imaging Cherenkov RICH. The midrapidity spectrometer has two time projection chambers TPC1 and TPC2, a dipole magnet D5, and a segmented scintillator time of flight wall TOFW. Multiplicity arrays close to the interaction point, and two zero degree calorimeters at ± 18m complete the set-up.

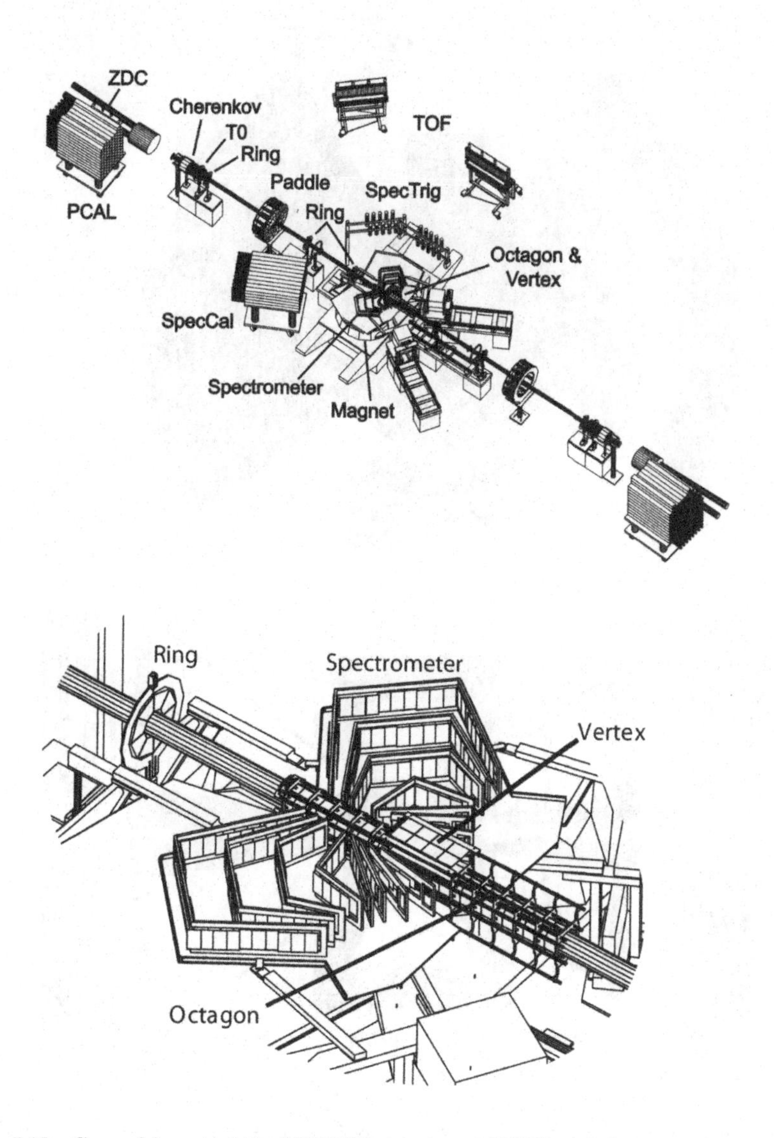

Fig. 5.13 General layout of the PHOBOS detector at RHIC, and, below, the enlarged view of the region around the beam collision point. The central part of PHOBOS consists of two-arm magnetic spectrometer with 2 Tesla field, and $|\eta| \leq 1$ acceptance, with multi-layer silicon detectors for charged particles tracking and momentum and dE/dx measurements. Two time-of-flight walls (TOF) improve particle identification, and additional detectors measure the total charged multiplicity and provide the trigger.

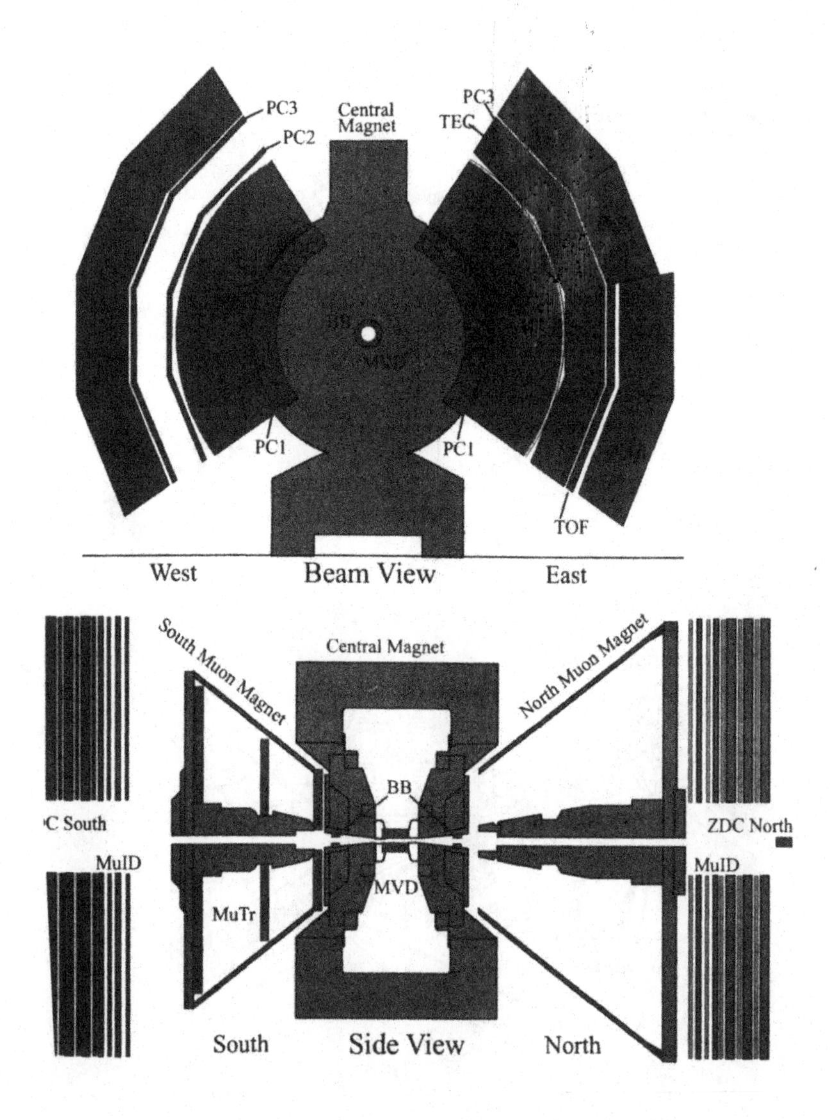

Fig. 5.14 Schematic picture of the PHENIX detector at RHIC (PHENIX = **P**ioneering **H**igh **E**nergy **N**uclear **I**nteractions **EX**periment). The PHENIX setup consists of four spectrometer arms: two around midrapidity (the central arms), and two at forward rapidity (the muon arms), and a set of global detectors. In each of the central arms charged particles are tracked by a drift chamber, and two or three layers of pixel pad chambers. Particle identification is provided by ring imaging Cherenkov counters, a time of flight scintillator wall, and two types of electromagnetic calorimeters (lead/scintillator and lead glass).

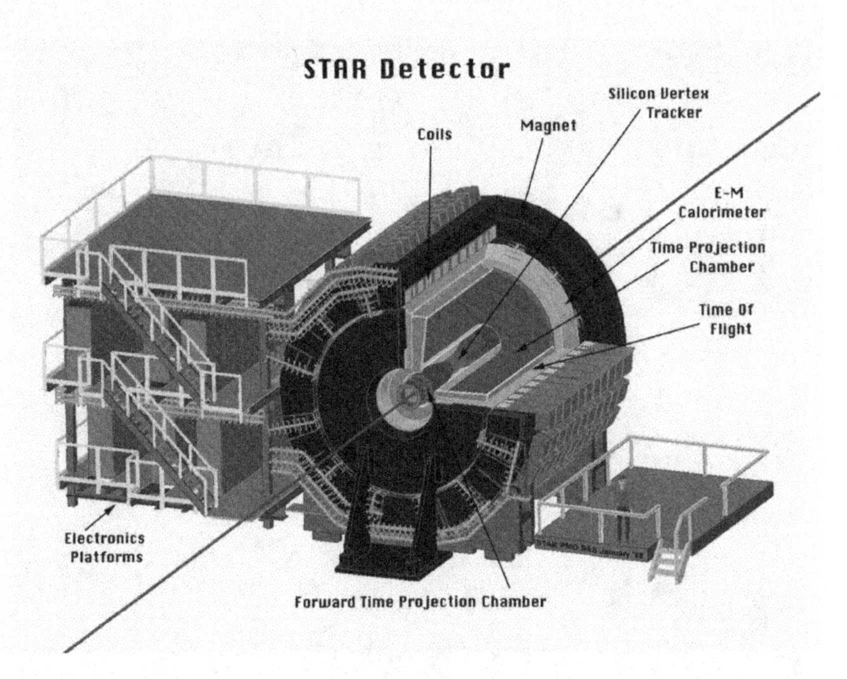

Fig. 5.15 Schematic picture of the STAR detector at RHIC (STAR = **S**olenoidal **T**racker **A**t **R**HIC). The setup consists of cylindrical detectors built around the interaction point, and placed in a magnetic field of 0.5 T, and of some forward detectors. The central detectors are: silicon vertex tracker, a big time projection chamber, ring imaging Cherenkov, time-of-flight counters, and electromagnetic calorimeter (the last three with only partial angular coverage). The forward detectors are: smaller time projection chambers, electromagnetic calorimeters, and zero degree calorimeters at large distance, all placed symmetrically on either side.

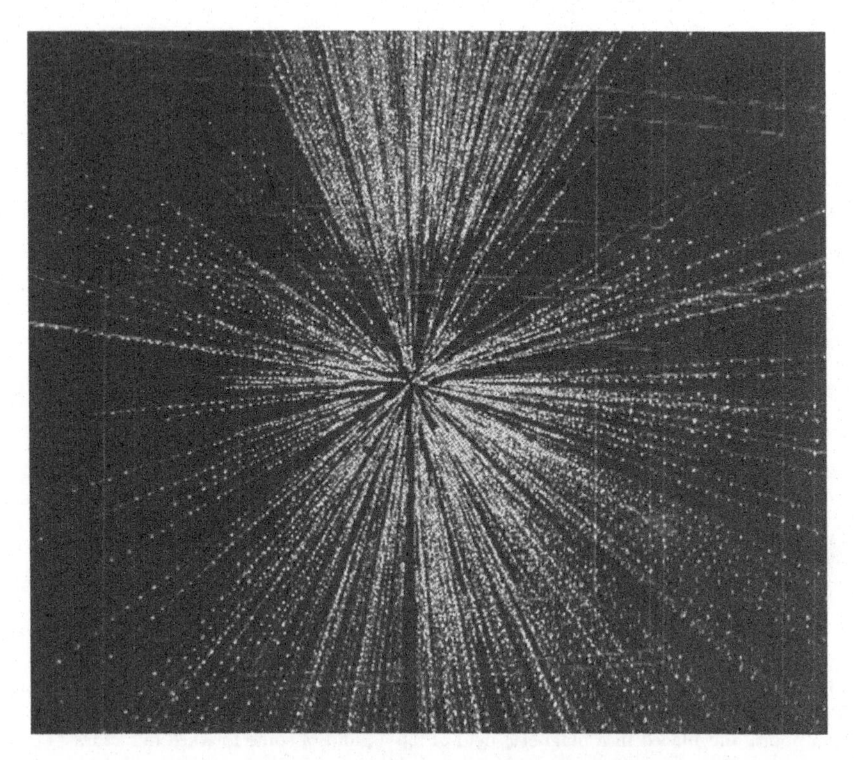

Fig. 5.16 Visualisation of tracks of charged particles emerging from a central Au+Au collision in the STAR detector at RHIC, emitted into a transverse slice around midrapidity.

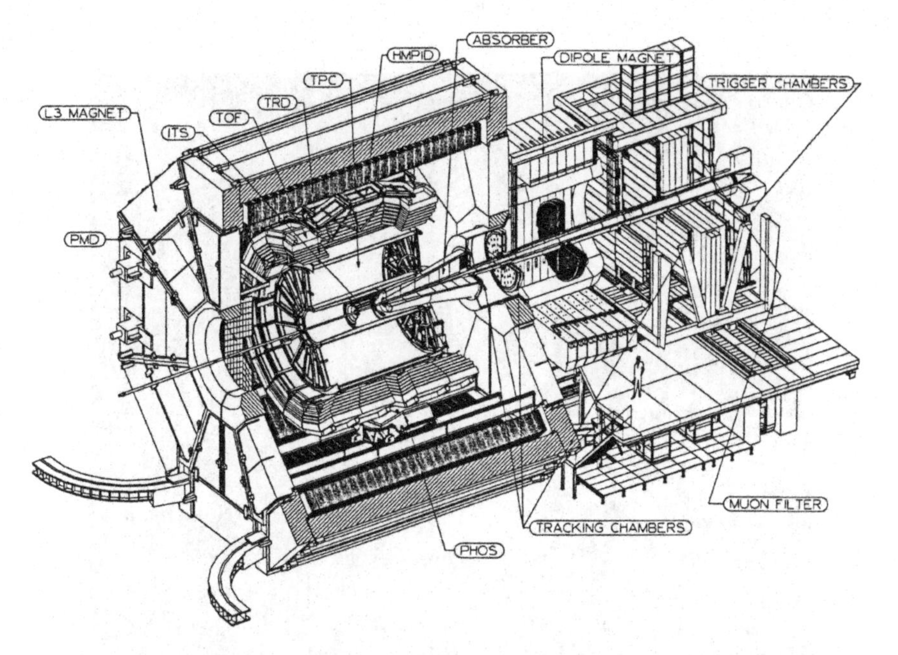

Fig. 5.17 Schematic picture of the ALICE detector for the LHC (ALICE = **A** **L**arge **I**on **C**ollider **E**xperiment). This is the largest detector built for investigating interactions of ultrarelativistic heavy ions. The main part of it is placed inside the big solenoidal magnet with an octogonal yoke and external dimensions 14m × 14m × 14m. This part consists of an inner tracking system ITS composed of six layers of high resolution silicon detectors, a cylindrical time projection chamber TPC serving as the main tracking detector, transition radiation detectors TRD, time of flight counters TOF, ring imaging Cherenkov detectors HMPID, and photon calorimeter PHOS. The forward muon arm, about 10 meters long, consists of a conical absorber, a large dipole magnet, and fourteen stations of multiwire tracking chambers, interlaid with a muon filter. Photon multiplicity array PMD at the back side, several small detectors for the trigger, and two zero-degree calorimeters located on either side at the distance of 160 meters from the interaction point, complete the set-up. The total weight of the ALICE detector is about 10,000 tons, of which 8,000 tons is the weight of the magnet.

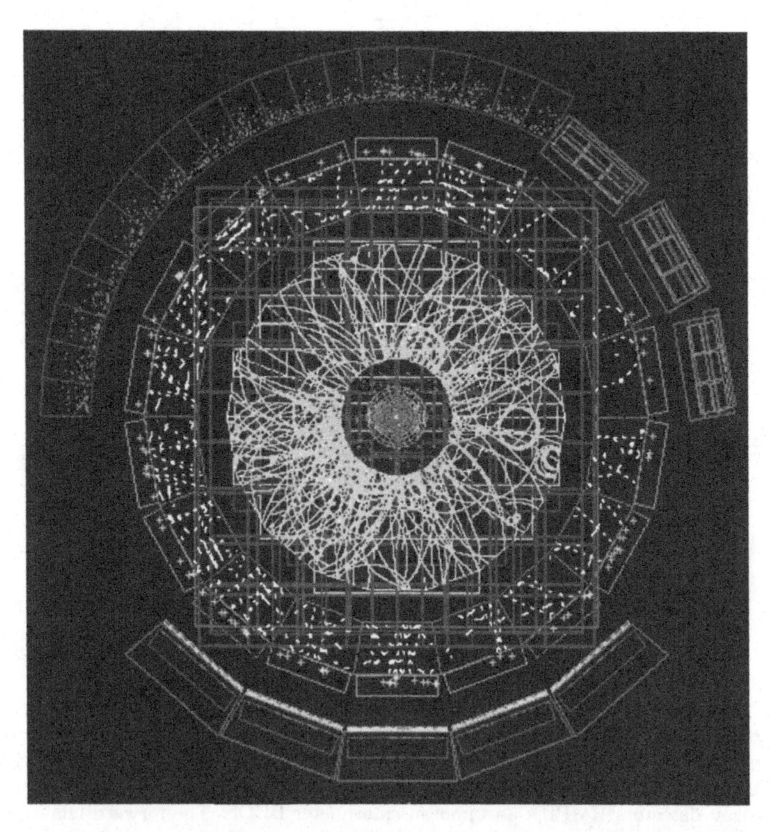

Fig. 5.18 Computer simulation of charged particle tracks emerging from a nuclear interaction in the ALICE detector at the LHC. Only a small fraction of the total number of tracks which for a central Pb+Pb collision is expected to be of the order of 10,000 could be shown in this picture.

Chapter 6

Cross Sections and Collision Geometry

6.1 Interaction cross sections

Let us consider a beam of N particles, or other objects, impinging upon a thin target. Due to interactions in the target, the intensity of the beam will be reduced by the amount

$$-dN(x) = N(x)\, n\, \sigma\, dx \tag{6.1}$$

Here n is the number of scattering centres per unit volume of the target, and $n\, dx$ is the number of scattering centres per unit area. The proportionality coefficient σ is called *the cross section* for interaction of incident particles with a given target. It has the meaning of the effective area of an elementary scattering centre.

Integration of Eq. (6.1) leads to the exponential absorption law, valid for a target of any thickness x

$$N(x) = N_0 \exp(-n\, \sigma x) \tag{6.2}$$

Another quantity, $\lambda = (n\, \sigma)^{-1}$, called *the mean interaction length*, is also being used to characterize the absorption properties of the material. In terms of this parameter Eq. (6.2) can be rewritten as

$$N(x) = N_0 \exp(-x/\lambda) \tag{6.3}$$

Cross section can be measured in a transmission experiment, by counting either unscattered particles, or, equivalently, the scattered ones, together with a measurement of the intensity of the incoming beam

$$\sigma = -\ln[N(x)/N_0]/n\, x \tag{6.4}$$

In a collision of two nuclei, due to the short range of nuclear forces, one can assume that nuclei begin to interact when their edges touch — see

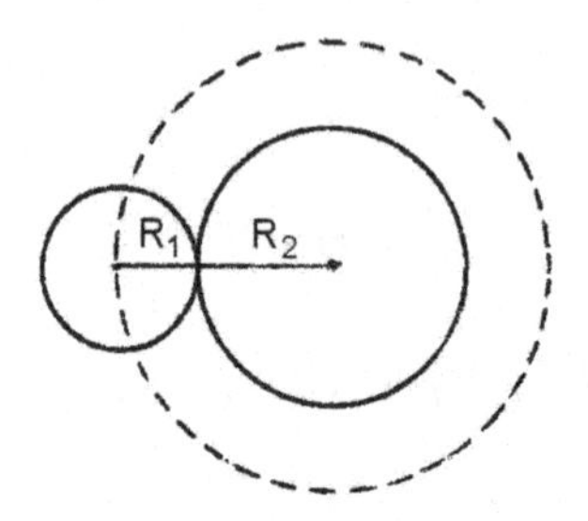

Fig. 6.1 Definition of the geometrical cross section.

Fig. 6.1. The interaction cross section which corresponds to this situation is called *the geometrical cross section*, and equals

$$\sigma_{\text{geom}} = \pi(R_1 + R_2)^2 = \pi r_0^2(A_1^{1/3} + A_2^{1/3})^2 \tag{6.5}$$

where the relation $R = r_0 A^{1/3}$ between nuclear radius R and the mass number A has been used. This very simple estimate turns out not to be far from reality. The formula found in early investigations of interactions of cosmic ray nuclei in emulsions, called Bradt and Peters formula [1], differs only slightly from Eq. (6.5), and reads

$$\sigma = \pi r_0^2(A_1^{1/3} + A_2^{1/3} - b)^2 \tag{6.6}$$

where b is the so-called *overlap parameter.* A least-squares two-parameter fit to a large number of cross sections measured at the Bevalac showed that while r_0 is insensitive to masses of colliding nuclei,[a] b turns out to be a monotonically decreasing function of A_{min}, the lighter of the colliding nuclei. The present-day experimental data can be well described by the improved formula

$$\sigma = \pi r_0^2[A_1^{1/3} + A_2^{1/3} - b_0(A_1^{-1/3} + A_2^{-1/3})]^2 \tag{6.7}$$

This parameterization is due to J. P. Vary, and the expressions $A_i^{-1/3}$ in the overlap term are related to effects of curvature of nuclear surfaces. Figure 6.2 shows the square root of the cross section as function of the ratio which characterizes the relative importance of the overlap term. Although this term includes contributions from both colliding nuclei, its value is dominated by the smaller nuclear mass, and becomes insensitive to changes in A_1 and A_2 when they are large. For heavy nuclei, the overlap term becomes

[a]This should have been expected as it reflects a constant density of matter in nuclei — see Chap. 3.

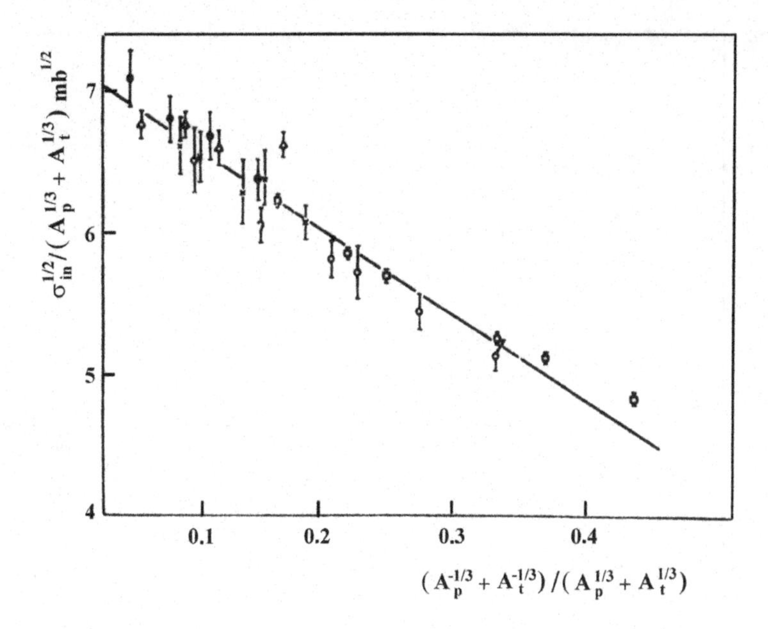

Fig. 6.2 Nuclear cross section data from the Bevalac and the Synchrophasotron plotted in a way showing the validity of Eq. (6.7) (from Ref. [2]).

small relative to $(A_1^{1/3} + A_2^{1/3})$, and the cross section approaches its geometrical limit, $\sigma_{\rm geom}$. The data at energies of a few GeV/nucleon, shown in Fig. 6.2, have been fitted with Eq. (6.7) with $r_0 = 1.3$ fm and $b_0 = 0.93$. It turns out that the same parameterization describes well also cross section data at AGS and SPS energies [3, 4] what means that interaction cross sections of relativistic nuclei do not change appreciably between $2A$ GeV and $200A$ GeV.

In collider experiments total cross section measurements would be extremely difficult, if at all possible, and cross section values calculated from the geometry of the colliding nuclei, using the Glauber model and the elementary nucleon-nucleon cross sections, are being used. They are believed to be accurate to a few percent.

For reference purposes, we show in Fig. 6.3 the total and elastic cross sections for proton–proton collisions plotted as functions of the energy of the collision. In the energy range of present-day accelerators $\sigma_{\rm pp}^{\rm tot} \approx 40$–$50$ mb, and rises slowly with increasing energy, while $\sigma_{\rm pp}^{\rm el} \approx 7$–$8$ mb, and is almost constant. Similar plots for other elementary cross sections can be found in Ref. [5].

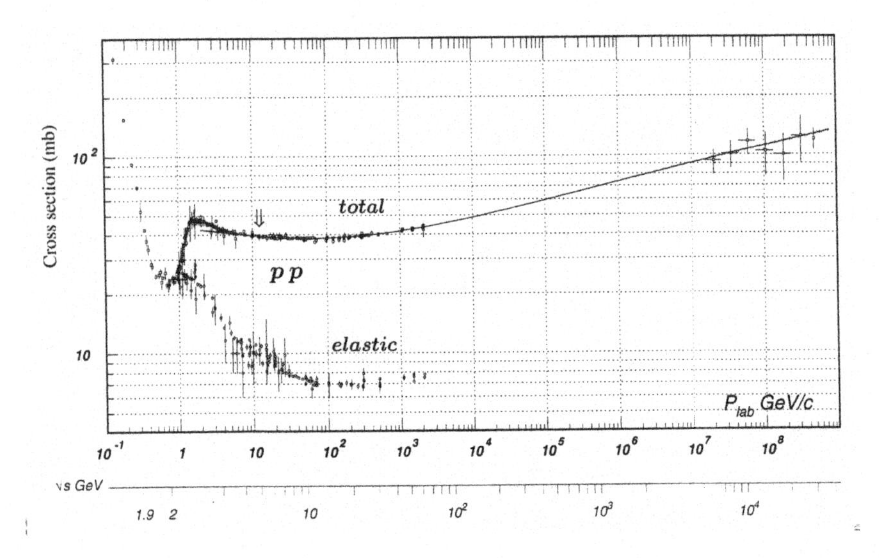

Fig. 6.3 Total and elastic proton–proton cross sections as functions of the incident momentum and of the c.m. energy of the collision (from Ref. [5]).

A few general remarks concerning hadronic cross sections will be of importance.

(i) On the basis of very general arguments involving unitarity, analyticity and crossing, Froissart showed that the total cross section for the strong interaction grows at most as fast as $\ln^2 E$ as $E \to \infty$. This is called the *Froissart bound* [6, 7].
(ii) Cross sections for particles and antiparticles converge as $E \to \infty$.
(iii) There is a simple relation between pion-nucleon and nucleon-nucleon cross sections: $\sigma_{\pi N} \approx \frac{2}{3}\sigma_{NN}$. This is usually quoted as a convincing evidence for the validity of the additive quark model, with the quark structure of baryons being (qqq) and that of mesons being $(q\bar{q})$.

6.2 Geometrical picture of the collision

The simple dependence of the total cross section of relativistic heavy ion reactions on the sizes of colliding nuclei suggests an important role of the collision geometry. This constitutes the basis of the *participant-spectator picture.* The colliding nuclei are assumed to move along straight-line trajectories, and only the geometrically overlapping parts of them interact, and what remains are the "spectators". This is schematically shown in Fig. 6.4 for a general case of a collision of two different nuclei, which was the case in many earlier experiments in which a relatively light projectile collided with a heavier target nucleus. As indicated in the lower panel of this Figure, in a central collision of such nuclei there is no projectile spectators. For equal mass nuclei, in a central collision there is no spectator matter whatsoever.

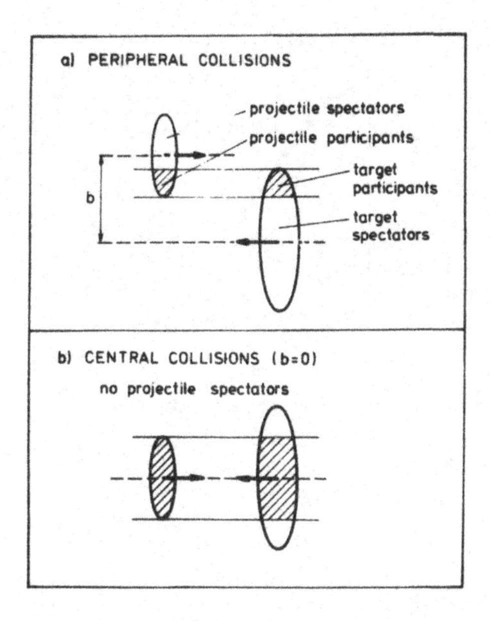

Fig. 6.4 Schematic diagram of the participant-spectator model.

An estimate of the collision time seems to justify this simple model. At relativistic energies the duration of the collision is very short as compared to the typical time scale for nuclear rearrangement or movement of nucleons within a nucleus. Thus nucleons finding themselves outside the overlap region should not experience any appreciable interaction. The "participant matter" is hot, while the "spectator matter" remains cold.

The centrality of a collision is determined by the impact parameter b (see Fig. 6.4), a quantity which, unfortunately, is not directly measurable. There are two complementary approaches to obtain an estimate of the centrality of a collision: one is based on a count of non-interacting (spectator) nucleons, the other uses some characteristics of produced particles (total multiplicity, transverse energy, etc.).

The participant-spectator picture allows for a simple calculation of the number of nucleons involved in a collision occuring with a given value of the impact parameter b. This is done using the Glauber model [8], assuming that nucleons in each nucleus are hard spheres distributed according to

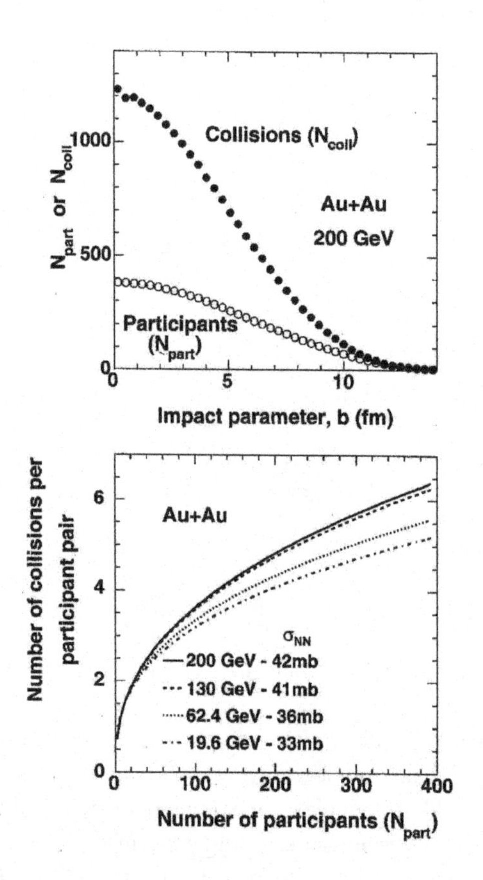

Fig. 6.5 Results of the Glauber model calculation for Au+Au collisions. The Saxon-Woods density distribution was assumed for the Au nucleus, with the mean radius $R =$ 6.38 fm, and the thickness of the surface layer d = 2.35 fm (see Chapter 3 for exact definitions of these parameters). Upper panel: number of binary collisions and number of nucleons-participants *vs.* the impact parameter b at $\sqrt{s_{NN}}$ = 200 GeV. Lower panel: number of collisions per participant pair obtained as the ratio of the number of collisions to the number of nucleons-participants at several energies. Results for different energies differ due to the variation of the nucleon-nucleon interaction cross section. Relevant values of this cross section are indicated in the figure (from Ref. [10]).

the nuclear density function (e.g. Saxon–Woods, see Chapter 3), and they move along parallel straight lines, interacting with nucleons from the other nucleus with cross sections known from elementary processes. Nucleons are treated as free particles, their internal motion and correlations in nuclei being neglected. When counting only the first collisions one obtains the number of nucleons-participants $N_{\rm part}$ (or "wounded" nucleons $N_{\rm wound}$ in the terminology of Białas *et al.* [9]), counting also subsequent collisions one obtains the total number of binary collisions $N_{\rm coll}$. These two quantities correspond to two limiting cases in the Glauber model, the first one being referred to as the "optical limit" of the model.

Results of the Glauber model calculations for Au+Au collisions are shown in Fig. 6.5. In the upper panel the quantities $N_{\rm part}$ and $N_{\rm coll}$ are plotted as functions of the impact parameter b for Au+Au collisions at $\sqrt{s_{NN}} = 200$ GeV, in the lower panel the ratio $N_{\rm coll}/N_{\rm part}$ is plotted *vs.* $N_{\rm part}$ for several values of the energy of the collision. This ratio gives the number of collisions suffered by each participant. For an unbiassed sample of collisions one obtains the mean value which should be equal to the quantity $\langle \nu \rangle$, traditionally used in the analysis of p+A collisions [11], and calculated from the relevant interaction cross sections as $\langle \nu \rangle = A\,\sigma_{\rm pp}/\sigma_{{\rm p}A}$.

It is worth noting that the Glauber calculation for A+A collisions yields the same value of $\langle \nu \rangle$ as for p+A collisions. As remarked in Ref. [12], this is an inherent feature of the geometrical picture of the collision, and of unbiassed samples in which each nucleon of one nucleus randomly scans the other nucleus, just as in a p+A collision. In Table 6.1 mean numbers of collisions per participant for unbiassed samples of collisions involving carbon and lead nuclei at 158A GeV are given as numerical examples.

Table 6.1 Mean number of collisions per participant nucleon in unbiassed samples of various nuclear collisions at 158A GeV (from Ref. [12]).

Colliding nuclei	$\langle \nu \rangle$
p+C (or C+C)	1.71 ± 0.05
p+Pb (or Pb+Pb)	3.75 ± 0.05

References

[1] H. C. Bradt and B. Peters, *Phys. Rev.* **77** (1950) 54.
[2] M. Kh. Anikina *et al.*, *Yad. Fiz.* **38** (1983) 149 [*Sov. J. Nucl. Phys.* **38** (1983)].
[3] A. Bamberger *et al.*, *Phys. Lett. B* **184** (1987) 271; *ibid* **205** (1988) 583.
[4] L. M. Barbier *et al.*, *Phys. Rev. Lett.* **60** (1988) 405.
[5] W. M. Yao *et al.* (Particle Data Group), *J. Phys. G: Nucl. Part. Phys.* **33** (2006) 1.
[6] M. Froissart, *Phys. Rev.* **123** (1961) 1053.
[7] A. Martin, *Nuovo Cim.* **42** (1966) 930.
[8] R. J. Glauber, *Nucl. Phys. A* **774** (2006) 3, and references therein.
[9] A. Białas, M. Błeszyński and W. Czyż, *Nucl. Phys. B* **111** (1976) 461.
[10] B. B. Back *et al.* (PHOBOS Collaboration), *Nucl. Phys. A* **757** (2005) 28.
[11] See *e.g.* W. Busza, *Acta Phys. Polon. B* **8** (1977) 333.
[12] A. Rybicki, INP Report No.1976/PH, Cracow (2006).

Chapter 7

Fragmentation Processes

In classical nuclear physics by "fragmentation" one means splitting of a nucleus into smaller parts, called nuclear fragments. The lightest fragments would be single nucleons, protons or neutrons, then come deuterons, helium nuclei, etc. A fragmentation process requires the energy transfer to the nucleus, and can be induced by electromagnetic forces — this is called *electromagnetic dissociation*, or by strong forces — this is called *nuclear fragmentation*. We will discuss these two processes separately, as they differ in many aspects, in particular in the cross-section behaviour with increasing incident energy. We will also address fragmentation in its extended meaning which is the notion used at very high energies.

7.1 Electromagnetic dissociation

In the process of electromagnetic dissociation (ED) some nucleons are separated from a nucleus as a result of alectromagnetic interaction with another nucleus. As the electromagnetic interaction is proportional to the square of the nuclear charge, Z, and falls steeply with the increasing four-momentum transfer q (interaction $\propto Z^2/q^4$), ED contributes significantly only to the "softest" processes, mainly to one-nucleon removal channels ${}^{A}Z \to {}^{A-1}Z$ and ${}^{A}Z \to {}^{A-1}(Z-1)$ on high-Z targets. At moderate energies also the two-nucleon removal has been observed, the probability of the separation of a larger number of nucleons being much smaller. Despite the limited number of reaction channels, the total cross-section for ED becomes very large at sufficiently high energies.

Electromagnetic dissociation of a nucleus is believed to proceed via its excitation by absorption of a virtual photon. The spectrum of virtual photons, $N_\gamma(E_\gamma)$, can be obtained by the Weizsäcker–Williams (W–W) method

[1]. Let us consider a physical system S (can be an atomic nucleus), and a particle with charge Q and velocity v passing by at a distance b (b is called *the impact parameter*) — Fig. 7.1. The moving particle generates the elec-

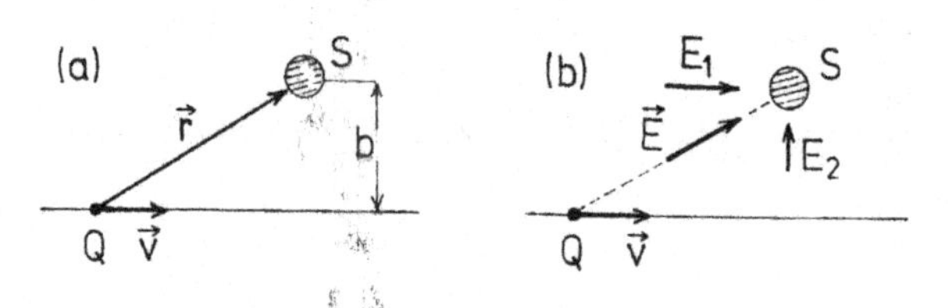

Fig. 7.1 Moving charged particle interacting with a system S (a) and its electric field vector (b).

tromagnetic field, the electric component of it can transmit some energy to the system S. The electric field vector $\vec{E}$ is along the radius vector $\vec{r}$, and its value and direction changes as the particle moves. It is convenient to split $\vec{E}$ into two components: longitudinal E_1 and transverse E_2.They are given by the following formulae:

$$E_1(t,b) = -Q\gamma vt/(b^2+\gamma^2v^2t^2)^{3/2} \tag{7.1}$$

$$E_2(t,b) = Q\gamma b/(b^2+\gamma^2v^2t^2)^{3/2} \tag{7.2}$$

and their behaviour with time is shown in Fig. 7.2. The W–W method

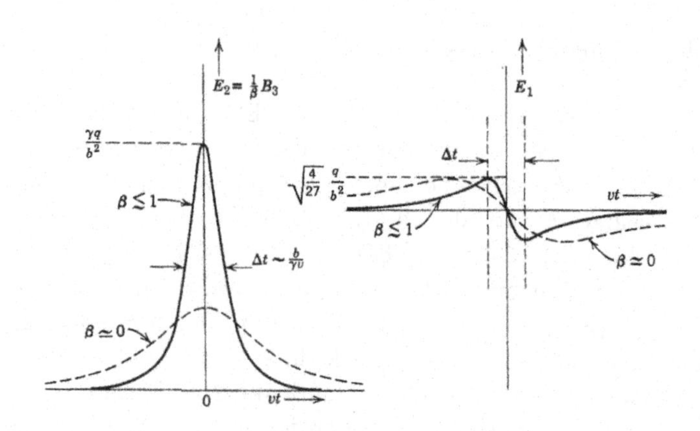

Fig. 7.2 Time dependence of the transverse (left) and longitudinal (right) components of the electric field generated by a moving relativistic charged particle (from Ref. [2]).

consists in replacing the field components E_1, E_2 with relevant radiation pulses, and considering the interaction of the system S with this radiation. Its frequency spectrum $dI(\omega,b)/d\omega$ can be obtained from $E(t,b)$ by the Fourier transform [2]. For each field component

$$\frac{dI_{1,2}(\omega,b)}{d\omega} = \frac{c}{2\pi}|E_{1,2}(\omega,b)|^2, \tag{7.3}$$

where

$$E(\omega,b) = \frac{1}{\sqrt{2\pi}}\int_{-\infty}^{+\infty} E(t,b)e^{i\omega t}dt \tag{7.4}$$

This function contains the dependence on the impact parameter b, and should be integrated over it

$$\frac{dI(\omega)}{d\omega} = 2\pi\int_{b_{\min}}^{\infty}\left(\frac{dI_1(\omega,b)}{d\omega} + \frac{dI_2(\omega,b)}{d\omega}\right)bdb \tag{7.5}$$

The result contains the leading term $\propto Z^2/E_\gamma$ multiplied by a rather complicated expression with Bessel functions. It depends on the minimum value of the impact parameter, $b_{\min}$. This can be obtained from the uncertainty principle: $b_{\min} \cong \hbar/Q_{\max}$ where $Q_{\max}$ is the largest allowed momentum transfer. This occurs in backward scattering and equals $Q_{\max} = 2mv$, and thus $b_{\min} = \hbar/2mv$ can be taken as approximate value for the lower limit of integration. If, however, this estimate of $b_{\min}$ turns out to be smaller than the nuclear radius R, one should take $b_{\min} = R$, as for $b < R$ the nuclear interaction would dominate. The spectrum of virtual photons contains mainly soft photons and falls down with increasing photon energy, extending to the limiting value of $\gamma\hbar c/b_{\min}$. Figure 7.3 shows the spectra of virtual photons, numerically calculated for the interaction of nuclear beams of different energies with a stationary uranium target. For each incident energy several curves are shown, corresponding to different modifications of the theory. It can be seen that at high energies they tend to converge.

In order to obtain the cross section for electromagnetic dissociation, the calculated virtual photon spectrum (one assumes that virtual photons interact like the real ones) should be folded with the relevant experimental photodissociation cross section $\sigma_\gamma(E_\gamma)$

$$\sigma_{\text{em}} = \int_0^{\infty} N_\gamma(E_\gamma)\sigma_\gamma(E_\gamma)dE_\gamma \tag{7.6}$$

The main contribution to the photodissociation cross section, as measured with real photons, comes from the giant dipole resonance, situated in the region of 20–25 MeV for light nuclei, and at about 14 MeV for heavy

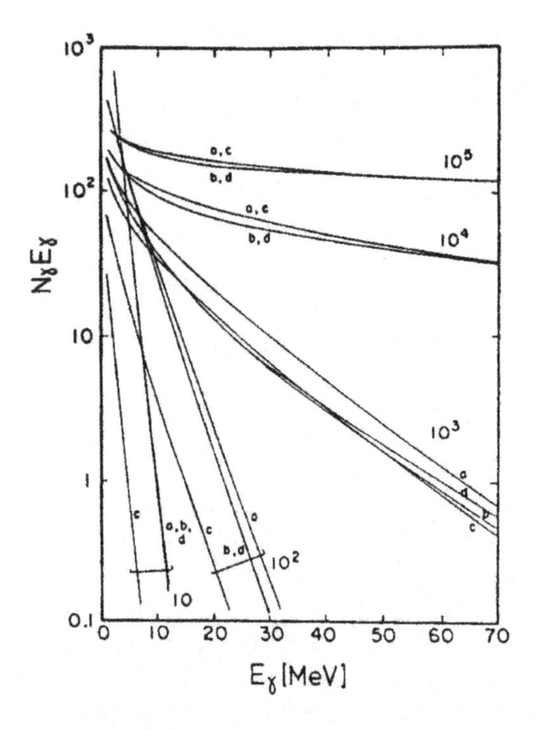

Fig. 7.3 Spectra of virtual photons calculated for charged particles of various energies (in MeV), interacting with uranium target (from Ref. [3]).

nuclei. Figure 7.4 shows, from top to bottom, and for two reactions: ^{18}O + U $\rightarrow ^{17}$O + X at 1.7A GeV (left) and ^{139}La + ^{197}Au $\rightarrow ^{196}$Au + X at 1.26A GeV (right), the virtual photon energy spectrum, $N_\gamma(E_\gamma)$, the experimental energy dependence of the relevant photodissociation cross section, $\sigma_\gamma(E_\gamma)$, and the product of the above two functions. The integral of the bottom curve yields the ED cross section. Figure 7.5 shows the experimental cross sections for a few photodissociation reaction channels of ^{32}S nuclei in a much wider energy range.

Observation of electromagnetic dissociation of relativistic nuclei was claimed in some cosmic ray studies [4], but a convincing identification of this process, with its characteristic $\propto Z^2$ dependence, was obtained at the Bevalac, in the reactions ^{12}C $\rightarrow ^{11}$C, ^{11}B and ^{16}O $\rightarrow ^{15}$O,^{15}N at 1.05A GeV and 2.1A GeV on various targets [5]. These studies were later extended to ^{18}O and ^{56}Fe nuclei [3, 6]. Further investigations, using the CR-39 plastic detectors instead of the magnetic spectrometer, were performed with ^{32}S

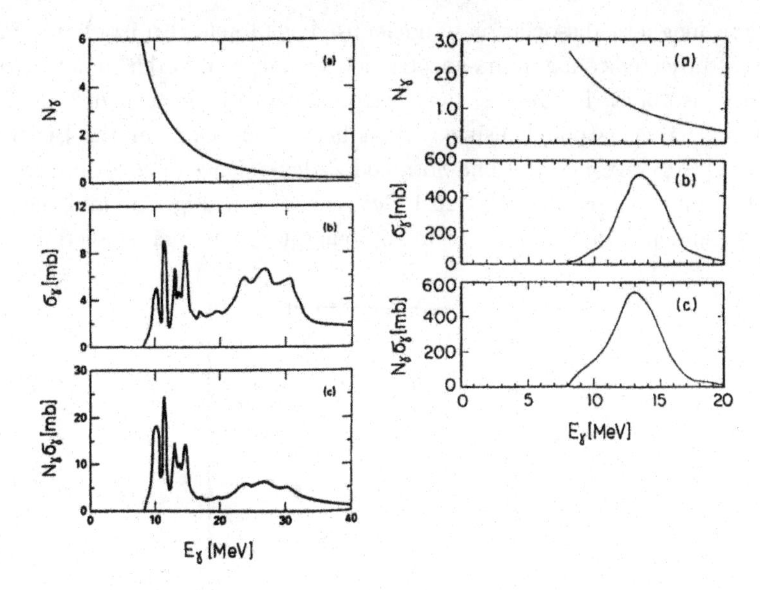

Fig. 7.4 Components necessary for calculation of the ED cross sections for the reactions $^{18}O + U \rightarrow {}^{17}O$ (left), and La$+{}^{197}Au \rightarrow {}^{196}Au$ (right) — details in text (from Refs. [3, 8]).

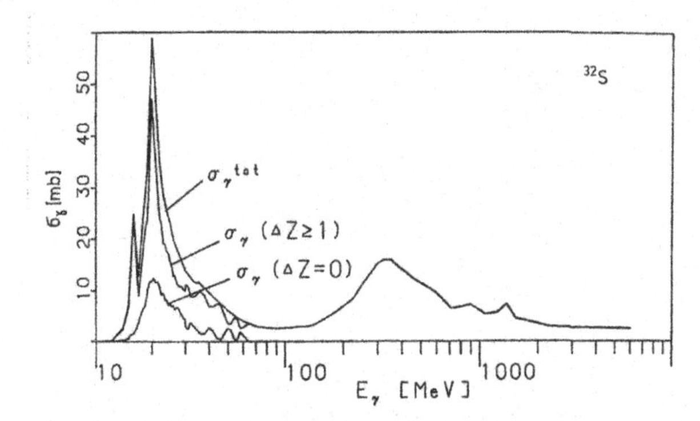

Fig. 7.5 Energy dependence of some photodissociation cross sections of ^{32}S nuclei (from Ref. [7]).

nuclei from the Bevalac, 14.5A GeV ^{28}Si nuclei from the Brookhaven AGS, and 60A GeV and 200A GeV ^{16}O and ^{32}S nuclei from the CERN SPS. For the latter case it was found that ED contributes to the removal of up to five nucleons from the sulphur nucleus [7].

A quite different technique, the activation method, was used for studies of electromagnetic dissociation of nuclei used as targets. An irradiated target was subjected to a gamma-spectrometric analysis in order to detect the produced isotopes. In this way the electromagnetic dissociation of ^{59}Co and ^{197}Au nuclei was studied, using various heavy ion beams at the Bevalac, AGS and SPS energies [9]. The obtained Z-dependence of electromagnetic dissociation cross sections at $1.26A$ GeV is shown in Fig. 7.6, and the energy dependence in Fig. 7.7. For high energies, where cross sections for

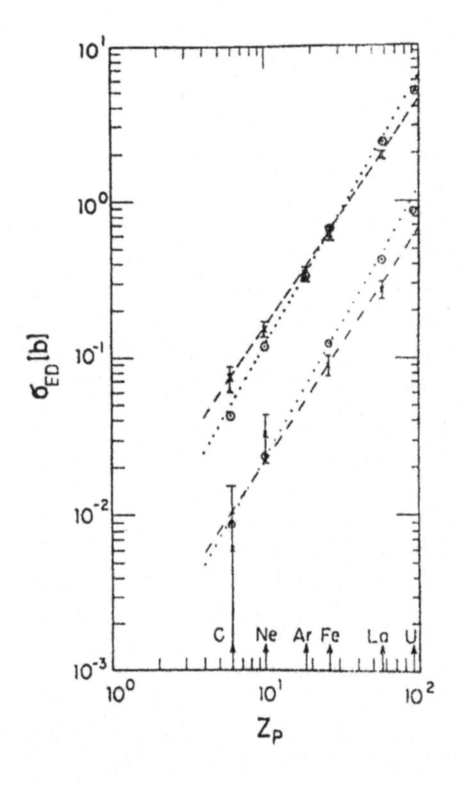

Fig. 7.6 Dependence of ED cross sections for the reactions $^{197}\mathrm{Au} \rightarrow ^{196}\mathrm{Au}$ (upper curves) and $^{59}\mathrm{Co} \rightarrow ^{58}\mathrm{Co}$ (lower curves) at $1.26A$ GeV on the charge Z of the projectile. Dashed lines are fitted to the data, while dotted ones, and round points, have been calculated (from Ref. [8]).

the competing nuclear fragmentation channels were not directly measured, they have been assumed to be the same as at lower energies (the hypothesis of "limiting fragmentation"). An overall approximate agreement with theoretical predictions for ED was found.

From Fig. 7.7 it can be seen that for heavy nuclei, and high energies,

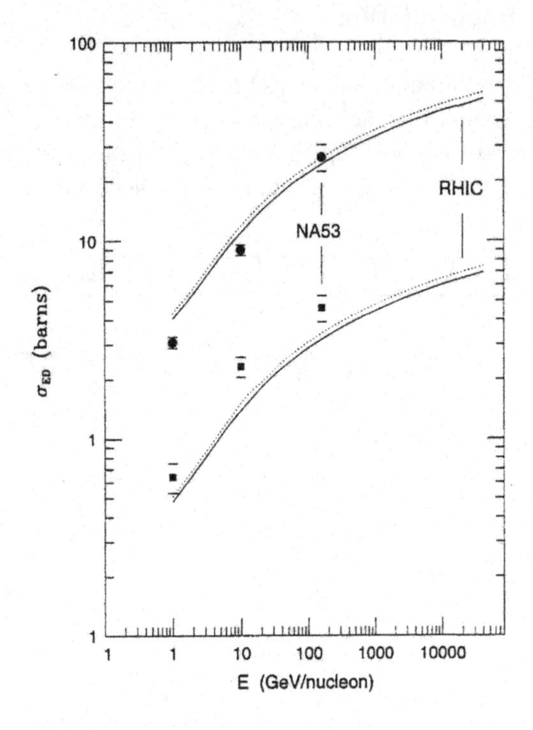

Fig. 7.7 Calculated energy dependence of cross sections for ED reactions in Au target. The upper and lower curves are for one- and two-neutron removal reactions, respectively, and the solid and dashed curves are for Au and Pb projectiles, respectively. Data points at 158A GeV, labeled NA53, are for Pb projectiles, those at lower energies are for Au projectiles (from Ref. [9]).

the cross sections for ED processes become fairly large. Extrapolation to the energies of RHIC, also shown in Fig. 7.7, indicates that for Au +Au collisions at $(100+100)A$ GeV the ED cross section is about 50 b, or almost ten times the geometrical cross section. This means that it is the ED, and not nuclear interactions, which limits the lifetime of the gold ion beams circulating in the RHIC collider. A similar estimate for Pb+Pb collisions in the LHC reaches above 100 b.

7.2 Nuclear fragmentation

Reaction channels characteristic for ED processes are olso open in nuclear fragmentation, as shown in the comparative diagram in Fig. 7.8, but many other final states are possible, up to a complete disintegration of the fragmenting nucleus. Nuclear fragmentation was studied in experiments which

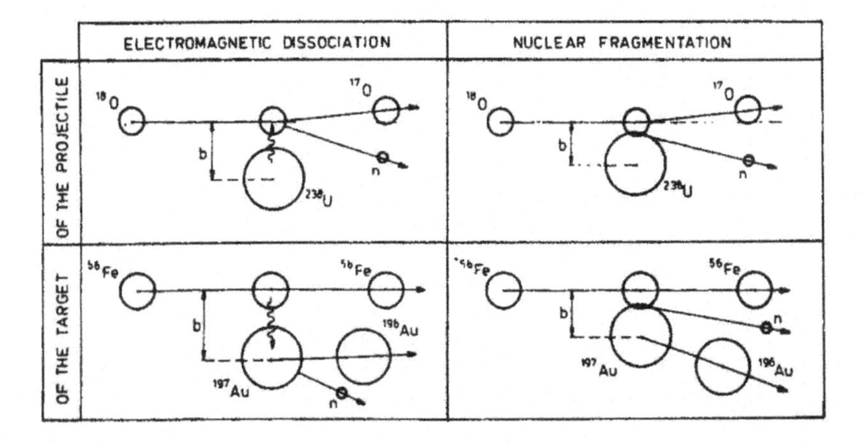

Fig. 7.8 Examples of single-nucleon separation processes by electromagnetic dissociation and by nuclear interactions.

have been mentioned above in connection to ED, but also in other ones. Main features of the fragmentation processes were established already in the early studies at the Bevalac [10].

It was found that the distributions of momentum components of the fragments in the rest frame of the projectile nucleus have a Gaussian shape with the width (std. dev.) between 50 and 200 MeV, depending only on the masses of the fragmenting nucleus, and of the fragment, and not on the target nucleus and the beam energy. The angular distribution of the fragments in the projectile rest frame is close to isotropy, and their momentum spectra indicate an effective temperature of 8–10 MeV, i.e. a very low excitation. The isotope production ratios are approximately target and energy independent. This suggests, as noticed by Feshbach [11], that the fragmentation process can be viewed as a decay of an excited nucleus, i.e. a delayed process which, according to Bohr's independence hypothesis, keeps little or no memory of the mechanism of its excitation. A simple theory formulated along these lines by Goldhaber [12] predicts for any momentum component of a fragment in the rest frame of the fragmenting nucleus a distribution of a Gaussian form with standard deviation

$$\sigma = \left(\frac{F(A-F)}{A-1} \langle p_i^2 \rangle \right)^{1/2} \quad (7.7)$$

where A and F are the mass numbers of the fragmenting nucleus and the fragment, correspondingly, and $\langle p_i^2 \rangle$ is the mean squared momentum of a nucleon in the fragmenting nucleus. This quantity is closely related to the Fermi momentum p_F. In the Fermi gas model of the nucleus which assumes a constant occupation density in momentum space, i.e. $dN/dp \propto p^2$ with a sharp cut-off at $p = p_F$, this relation is $\langle p_i^2 \rangle = \frac{3}{5} p_F^2$ [12]. However, the Gaussian shape of the momentum component distributions found in experiment points towards a Gaussian momentum distribution of nucleons in the nucleus: $dN/dp \propto p^2 exp(-p^2/2\sigma^2)$. Figure 7.9 shows the results on fragmentation of ^{16}O projectiles at 200A GeV [13]. The distribution has

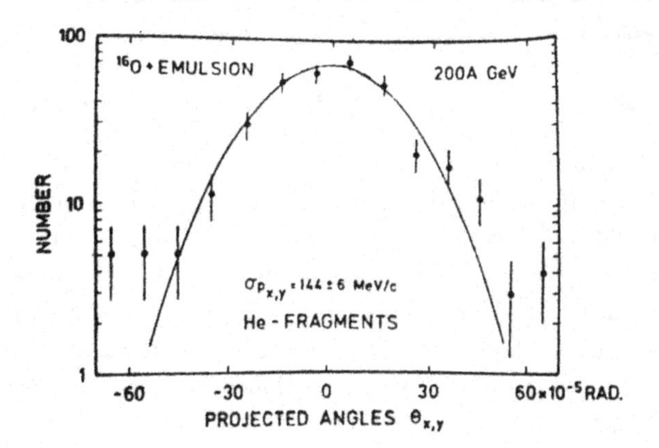

Fig. 7.9 Distribution of transverse momentum components in fragmentation of 200A GeV ^{16}O nuclei in emulsion (from Ref. [13]).

the width of 144 ± 6 MeV/c, similar to that found at low energies (137 ± 2 MeV/c at 2A GeV [14]), and a close-to-Gaussian form.

The mean angle of emission of a fragment is given by the ratio of its transverse momentum which is of the order of the Fermi momentum p_F to the longitudinal momentum. Fragments from a fast moving nucleus of mass A and total momentum AP[a] have velocities close to that of the projectile, or longitudinal momenta $p_L \approx FP$ with F denoting the mass of a fragment, and thus heavier fragments are emitted into a narrower angular

[a] Here P denotes the momentum per nucleon.

cone. Figure 7.10 displays the pattern of fragmentation of ^{208}Pb nuclei at 158A GeV, as it appears on the plane perpendicular to the beam direction at the distance of 25 m from target, after traversing the integrated magnetic field of 7.8 Tm of the NA49 spectrometer at CERN [15]. The magnetic

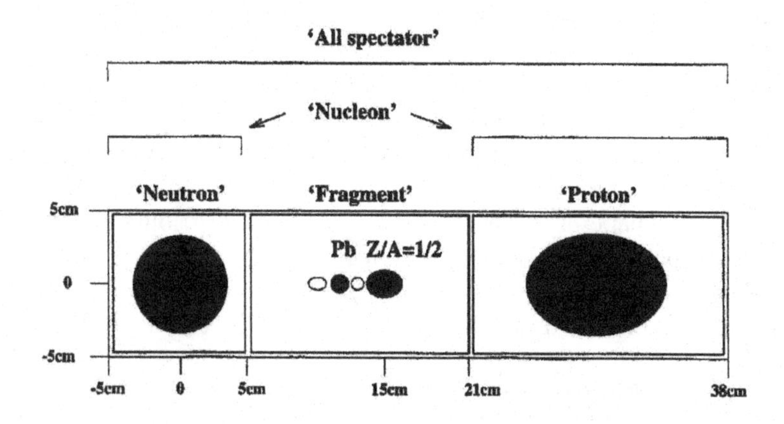

Fig. 7.10 Pattern of fragmentation of 158A GeV ^{208}Pb nuclei in the CERN NA49 spectrometer (from Ref. [15]).

field is vertical and causes deflection of charged particles in the horizontal plane, the deflection depending on the charge-to-mass ratio, Z/A, with some spread due to the Fermi momentum. This deflection is 11.9 cm for the ^{208}Pb nuclei of the incident beam ($Z/A = 0.39$), 15.1 cm for fragments with $Z/A = 0.5$, and 30.3 cm for protons. Broadening of the distribution for each species is due to the Fermi motion, which also causes oval shapes for charged fragments.

The isotopic spectra of fragments are also interesting. Figure 7.11 shows the charge spectrum of fragments from ^{12}C nuclei at 3.66A GeV, recorded by the Cherenkov detector in an experiment at JINR, Dubna [16]. The spectrometer allowed to obtain the full isotopic spectrum of fragments — examples are given in Fig. 7.12. A very high yield of helium is visible already in the charge spectrum of fragments, the isotopic analysis shows that this is almost exclusively ^{4}He. This points out towards the alpha-cluster substructure of the ^{12}C nucleus. Among lithium isotopes, the ^{6}Li is the most abundant.

Figure 7.13 shows the charge spectrum of the $Z/A = 1/2$ fragments obtained from the 158A GeV lead beam from the CERN SPS. The beam hitted a 10 mm thick carbon target, and fragmentation products were analyzed on

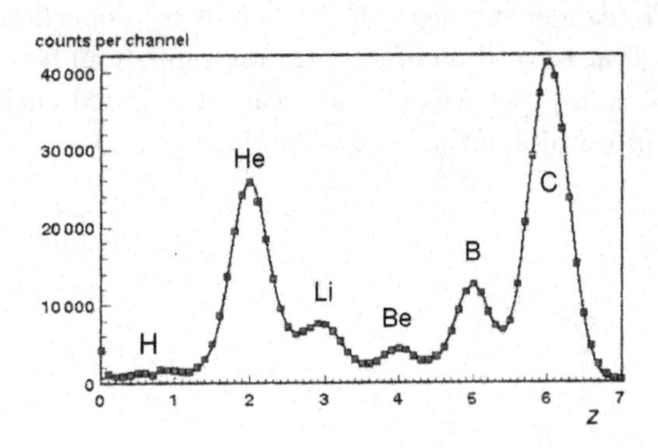

Fig. 7.11 Charge spectrum of fragments from 3.66A GeV ^{12}C nuclei (from Ref. [16]).

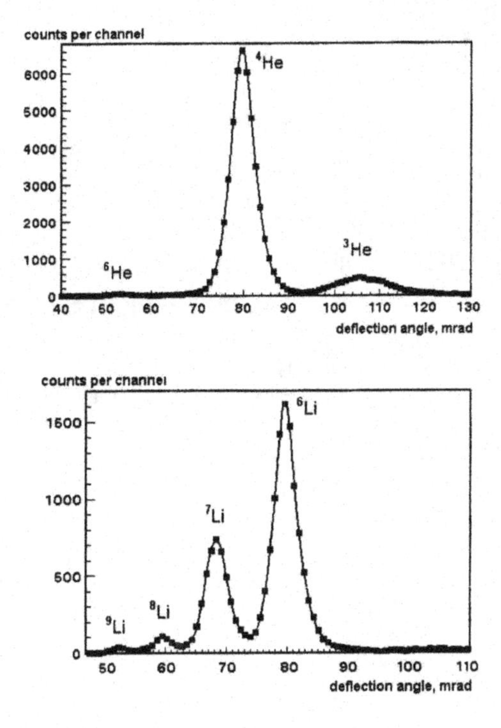

Fig. 7.12 Isotopic spectrum of helium and lithium fragments from 3.66A GeV ^{12}C nuclei (from Ref. [16]).

the basis of the pulse height in a scintillation counter, combined with the energy loss dE/dx measurement in six multiwire proportional chambers. In this way a clear separation of nuclear fragments from boron ($Z = 5$) to phosphorus ($Z = 15$) was achieved, and various selected nuclei could have been used as projectiles in the NA49 experiment.

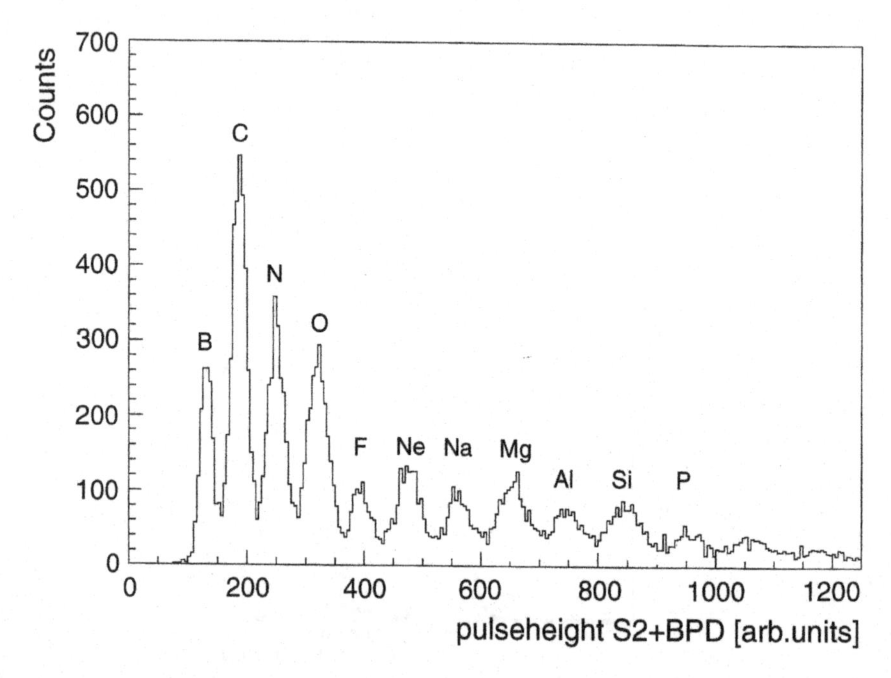

Fig. 7.13 The 158A GeV lead beam fragmentation spectrum from a 10 mm thick carbon target in a secondary beam line set to select $Z/A = 1/2$ fragments (from Ref. [17]).

7.3 Fragmentation in its extended meaning

At very high energies, a collision of two hadrons leads to particle emission from three regions: that of the projectile, that of the target, and a central one, as schematically shown in Fig. 7.14. Particles emitted from outer regions are called products of the projectile (or target) fragmentation. This is based on the parton model in which an excited hadron can "fragment" into partons, quarks and gluons, and quarks can recombine in new hadrons. Similarly, in high energy nuclear collisions all hadrons emitted into the rapidity interval close to the rapidity of the incident nucleus, and not only those carrying baryonic number, are called *projectile fragments.* This is the extended meaning of "fragmentation", which in lower-energy nuclear physics just means a decomposition of a nucleus into lighter nuclear species, or single nucleons. The width of the "fragmentation region" in rapidity is about two units. Thus one can expect fragmentation regions to be well separated from the region of "central" particle production only at very high energies, where the total kinematically available rapidity interval is much wider than four units.

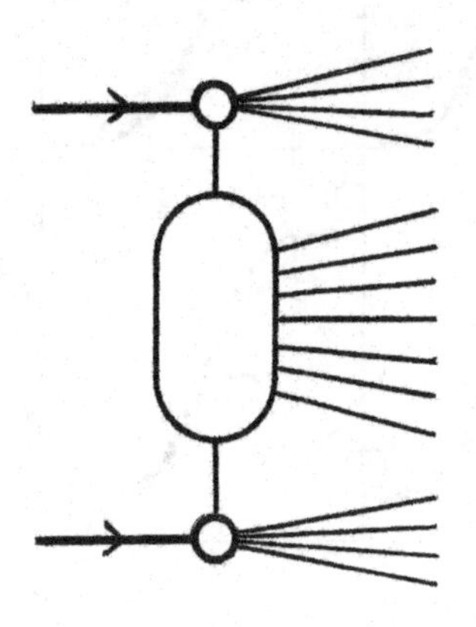

Fig. 7.14 Schematic diagram of particle production in a high energy hadronic collision

Figure 7.15 shows pseudorapidity distributions of charged particles produced in Au+Au collisions at several energies, and for two values of centrality, as recorded by the PHOBOS experiment at RHIC [18]. The data have been shifted to the rest frame of each of the colliding gold nuclei by plotting them versus $\eta' = \eta \pm y_{\text{beam}}$. Overlapping of distributions at, correspondingly, left or right edge can be seen at all energies, and for both centralities. In the far right panels of Fig. 7.15 data for positive and negative η have been averaged, in order to better show this feature, which is called the "limiting fragmentation".[b] No well separated central region has been observed up to top RHIC energy of $(100 + 100)A$ GeV, where $y_{\max} = 5.4$.

[b] Also called "longitudinal scaling" as rapidity is a longitudinal variable.

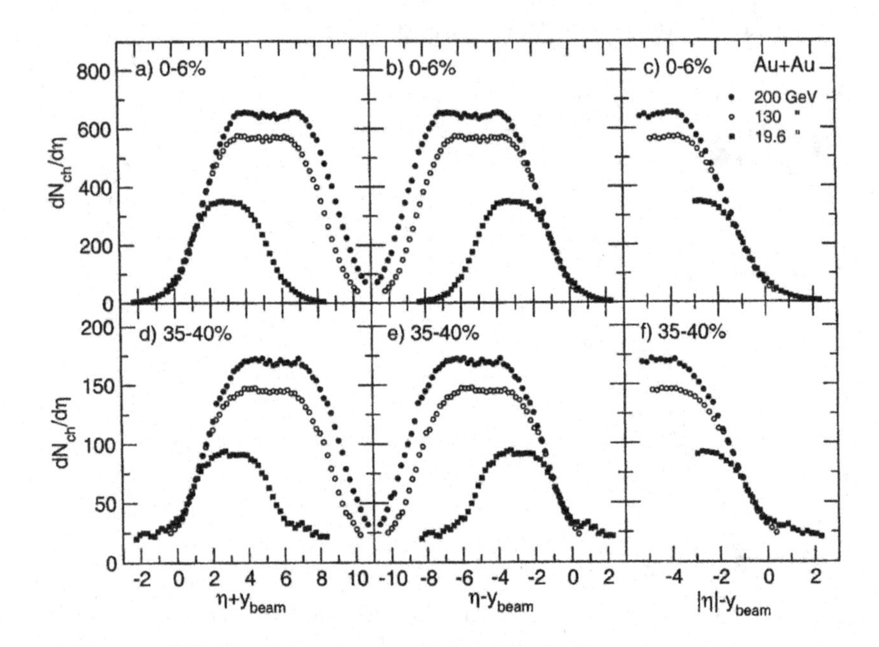

Fig. 7.15 Pseudorapidity distributions of charged particles produced in Au + Au collisions at three energies and two centralities, plotted so as to show the "limiting fragmentation"feature (from Ref. [18]). For more details see text.

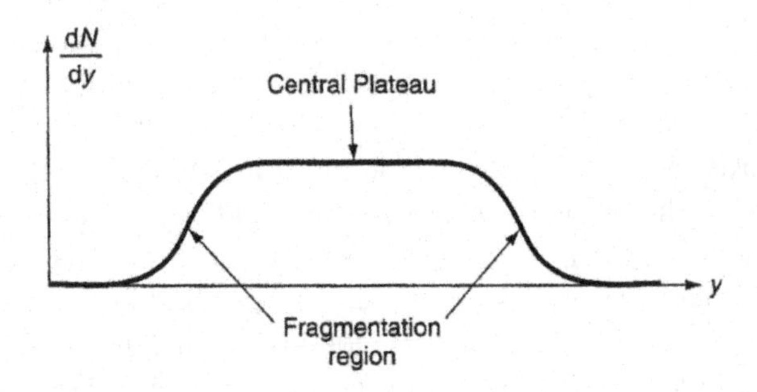

Fig. 7.16 Shape of secondary particles rapidity distribution expected at LHC energies.

For Pb+Pb collisions in the LHC at $(2.75 + 2.75)A$ TeV $y_{\max} = 8.7$ and a central plateau should develop, as shown in Fig. 7.16.

References

[1] C. F. v. Weizsäcker, *Z. Phys.* **88** (1934) 612; E. J. Williams, *Phys. Rev.* **45** (1934) 729.
[2] J. D. Jackson, *Classical Electrodynamics*, 2nd edn., Sec. 11.10 (John Wiley, New York, 1975).
[3] D. L. Olson *et al.*, *Phys. Rev. C* **24** (1981) 1529.
[4] K. Rybicki, *Nuovo Cim.* **49** (1967) 203; X. Artru and G. B. Yodh, *Phys. Lett. B* **40** (1972) 43.
[5] H. H. Heckman and P. J. Lindstrom, *Phys. Rev. Lett.* **37** (1976) 56.
[6] G. D. Westfall *et al.*, *Phys. Rev. C* **19** (1979) 1309.
[7] C. Brechtmann and W. Heinrich, *Z. Phys. A* **331** (1988) 463.
[8] J. C. Hill *et al.*, *Phys. Rev. C* **38** (1988) 1722.
[9] J. C. Hill *et al.*, *Nucl. Phys. A* **661** (1999) 313c.
[10] H. H. Heckman *et al.*, *Phys. Rev. Lett.* **28** (1972) 926.
[11] H. Feshbach and K. Huang, *Phys. Lett. B* **47** (1973) 300.
[12] A. S. Goldhaber, *Phys. Lett. B* **53** (1974) 306.
[13] I. Otterlund, *Z. Phys. C* **38** (1988) 65.
[14] D. E. Greiner *et al.*, *Phys. Rev. Lett.* **35** (1975) 152.
[15] H. Appelshäuser *et al.*, *Eur. Phys. J. A* **2** (1998) 383.
[16] A. Korejwo *et al.*, *J. Phys. G: Nucl. Part. Phys.* **26** (2000) 1171.
[17] J. Bächler *et al.* (NA49 Collaboration), Status and future programme of the NA49 experiment, CERN/SPSLC/P264 Add. 3 (13 January 1999).
[18] B. B. Back *et al.*, *Phys. Rev. Lett.* **91** (2003) 052303.

Chapter 8

Multiplicities and Relative Abundances of Secondary Particles

8.1 Mean multiplicities

In collisions of relativistic nuclei, like in any hadronic interactions at high energy, various secondary particles are being produced. Their number depends on the energy of the collision, and on the "centrality", or degree of overlap of colliding nuclei, which determines the number of nucleons participating in the collision (see Chapter 6). For "central" collisions of heavy nuclei at high energies the total multiplicity of secondary particles can be very large. Table 8.1 shows the mean numbers of charged secondaries from central S+S[a] and Pb+Pb collisions at the SPS [1, 2], and Au+Au collisions at three energies [3]. The total number of participating nucleons in central Au+Au collisions is estimated to be about 350. A proportionality of

Table 8.1 Mean number of charged secondaries in central collisions of various nuclei at several energies (from Refs. [1–3]).

Reaction	$\sqrt{s_{NN}}$, GeV	$\langle N_{\rm ch}\rangle$
S+S	19.6	228 ± 6
Pb+Pb	17.3	~ 1600
Au+Au	19.6	1744 ± 131
Au+Au	130	4376 ± 219
Au+Au	200	5290 ± 264

the number of produced particles to the number of participating nucleons

[a]For S+S collisions the mean multiplicity of charged secondaries has been calculated from the published data on negatively charged particles, $\langle N_-\rangle$. As sulphur is an isotopically symmetric nucleus with $Z = N = 16$, $\langle N_{\rm ch}\rangle = 2\langle N_-\rangle + 2 \cdot 16$.

was found already in early experiments at the Bevalac [5, 6] and in Dubna [7]. Figure 8.1 shows the multiplicity of negatively charged secondaries (at these energies they all are pions) produced in La+La collisions at energies from $0.53A$ GeV to $1.35A$ GeV as a function of the number of participating nucleons. The above mentioned proportionality is clearly seen. A similar situation is observed at higher energies — see below.

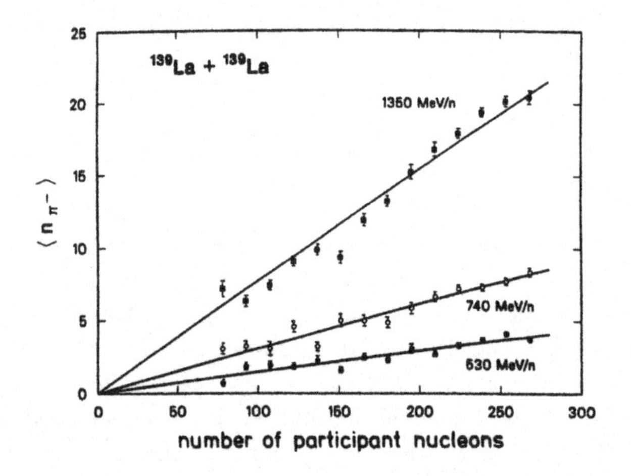

Fig. 8.1 Mean pion multiplicity as a function of the number of participating nucleons in La+La collisions at three incident laboratory energies. The straight lines are fitted to the data points (from Ref. [6]).

Instead of the number of participating nucleons, $N_{\rm part}$, the number of participating nucleon pairs, $N_{\rm part}/2$, is used more often to show scaling properties of various characteristics of nuclear collisions. Figure 8.2 shows the total multiplicity of charged secondary particles per participant pair as a function of the number of participants in Au+Au collisions at three energies, in d+Au collisions at $\sqrt{s} = 200A$ GeV, and in p+$\bar{\rm p}$ collisions at $\sqrt{s} = 200$ GeV. Two interesting features of the data are visible. First, similarly to earlier observations, for a given colliding system the multiplicity of secondary particles is proportional to the number of participants over a wide range of its values. Second, the multiplicity per participant pair in p+$\bar{\rm p}$ or d+Au collisions is significantly lower than that in Au+Au collisions at the same energy per nucleon. This is clearly seen in Fig. 8.3 in which multiplicities per nucleon pair are compared for various colliding systems. For hadron-nucleus and d+Au collisions they agree with multiplicities in

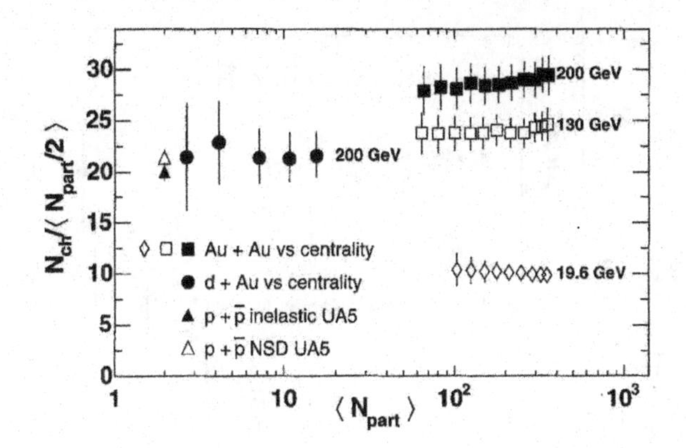

Fig. 8.2 Total charged particle multiplicity per participant pair as a function of the number of participants for Au+Au collisions at $\sqrt{s_{NN}} = 19.6, 130$ and 200 GeV, and also for d+Au and p+p̄ collisions at $\sqrt{s_{NN}} = 200$ GeV (from Ref. [8]).

p+p̄ collisions at the same incident energy, while multiplicities in Au+Au collisions agree with those in p+p̄ collisions taken at twice the incident energy. It is well known that in a p+p(p̄) collision only about one half of the available energy is used for particle production, the other half is retained by the "leading nucleons" (one speaks about *the inelasticity coefficient* being close to 0.5). It is most interesting that nucleons bound in nuclei seem to be more effective in particle production.

The multiplicity density, dN_{ch}/dy, or $dN_{\text{ch}}/d\eta$, near midrapidity, is often taken as a convenient measure of the total multiplicity. In Fig. 8.4 this quantity is plotted as a function of the collision energy. The straight line fitted to data points plotted in a semi-log scale shows that this dependence is logarithmic:

$$dN_{\text{ch}}/d\eta \propto \ln \sqrt{s_{NN}} \tag{8.1}$$

In Fig. 8.5 the multiplicity density near midrapidity is again displayed as a function of energy, for various high energy reactions. One can see that, similarly to the data on total multiplicities which have been discussed earlier, values of the multiplicity density near midrapidity for p+p(p̄) collisions are systematically lower than those for nuclear collisions at the same per-nucleon energy. A striking observation is that multiplicities in e^+e^- collisions follow the same dependence as those for collisions of relativistic nuclei.

The total multiplicity of charged secondaries, N_{ch}, approximately fol-

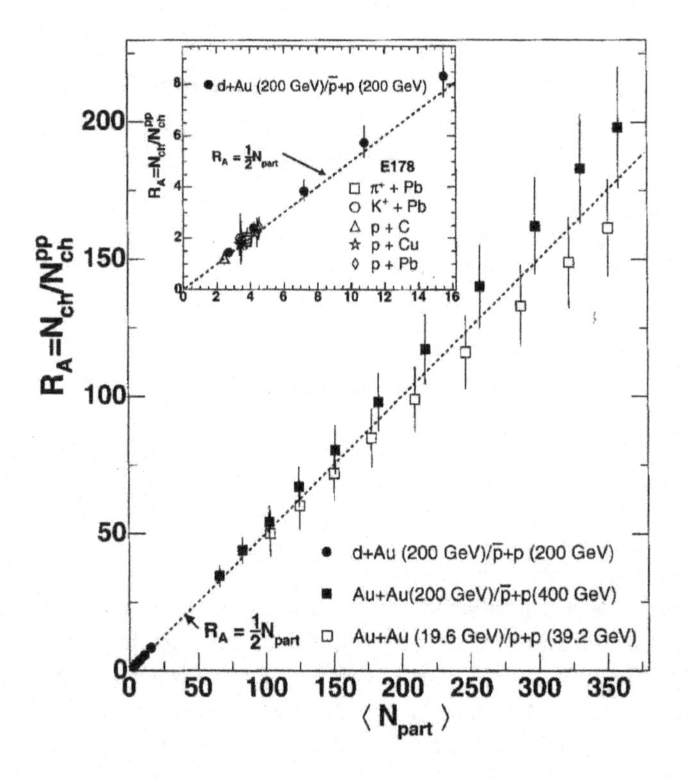

Fig. 8.3 Ratios of total multiplicities of charged secondaries in various hadron-nucleus and nucleus-nucleus collisions to the multiplicities in proton(antiproton)-proton interactions, plotted against the number of participating nucleons. For interactions induced by mesons, protons, and deuterons the proton(antiproton)-proton data are taken at the same c.m.energy, while for Au+Au collisions the proton(antiproton)-proton data are taken at twice the c.m.energy (from Ref. [8])

lows a logarithmic-squared dependence on the collision energy

$$N_{\rm ch} \propto \ln^2 \sqrt{s_{NN}} \tag{8.2}$$

This is a combined result of Eq. (8.1) and of the widening of the total multiplicity distribution in rapidity with increasing energy. This distribution, $dN_{\rm ch}/dy$, has an approximately trapezoidal shape, and its width increases as $y_{\rm beam}$ which at high energies is proportional to $\ln \sqrt{s_{NN}}$.

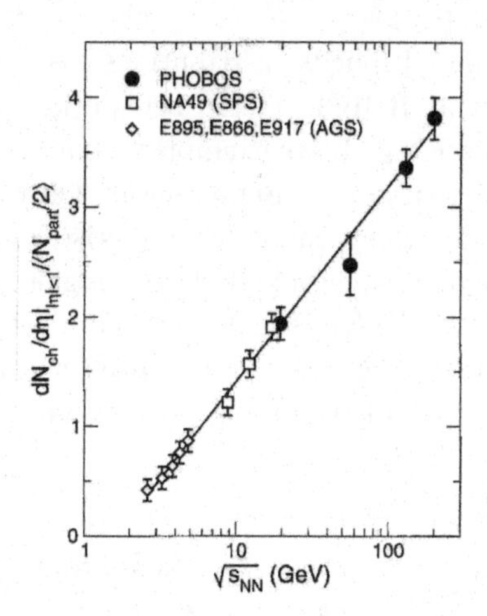

Fig. 8.4 Density of charged secondary particles near midrapidity (in the pseudorapidity interval $-1 < \eta < 1$) divided by the number of participating nucleon pairs, plotted as a function of the c.m.energy of the collision, for central Au+Au at AGS and RHIC, and for central Pb+Pb collisions at the SPS (from Ref. [8]).

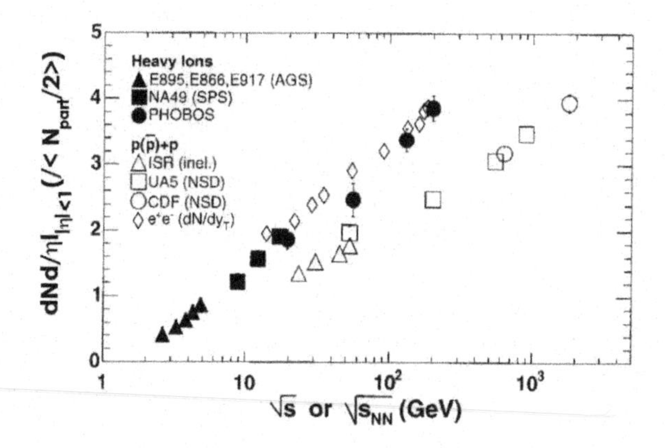

Fig. 8.5 Densities of charged secondary particles near midrapidity (in the pseudorapidity interval $-1 < \eta < 1$) divided by the number of participating nucleon pairs, plotted as a function of the c.m.energy of the collision, for central nucleus-nucleus collisions (data from AGS, SPS, and RHIC), proton(antiproton)-proton collisions, and e^+e^- collisions (from Ref. [8]).

8.2 Multiplicity distributions

The shape of the total multiplicity distributions has been studied in several experiments (but not at RHIC). These distributions are wide, especially for heavy nuclei. Figure 8.6 shows exemplary multiplicity distributions of secondary charged particles from interactions of $200A$ GeV ^{16}O nuclei with various nuclear targets. They have a characteristic shape: a peak at low multiplicities, a more or less developed plateau region, and a steep fall. For a given target they obey the KNO scaling [9], i.e. data at different energies look the same if plotted versus the "normalized multiplicity", $N/\langle N\rangle$ [4]. The shape of multiplicity distributions is determined by the collision ge-

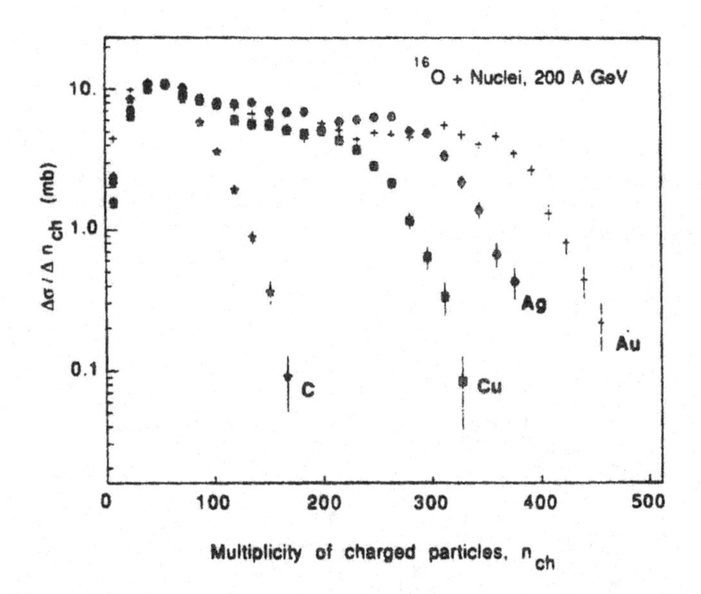

Fig. 8.6 Multiplicity distributions of secondary charged particles from ^{16}O + (C, Cu, Ag, Au) collisions at $200A$ GeV, measured in a wide angular interval ($-1.7 < \eta < 4.2$) (from Ref. [10]).

ometry. Large cross sections at low multiplicities correspond to peripheral collisions with large impact parameters, this is followed by an intermediate region of gradually increasing overlap of the colliding nuclei, and finally by a rapid decrease at their full overlap. Figure 8.7 shows multiplicity distributions of negatively charged hadrons from "minimum bias" (i.e. recorded without any selection) Au+Au collisions at $\sqrt{s_{NN}} = 130$ GeV. Particles come from a narrow angular interval near midrapidity (the pseudorapidity interval $-0.5 < \eta < 0.5$), and have transverse momenta $p_T > 100$ MeV/c. One can see that the overall shape of the distribution resembles very much

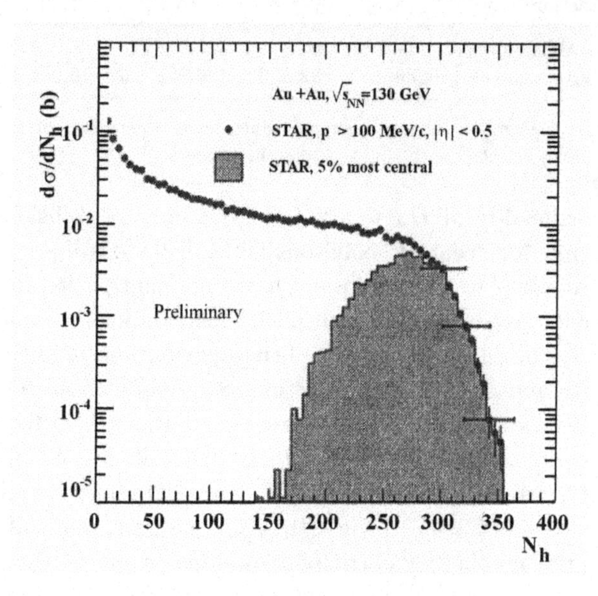

Fig. 8.7 Multiplicity distribution of negatively charged hadrons emitted into the pseudorapidity interval $-0.5 < \eta < 0.5$, and having $p_T > 100$ MeV/c, from Au+Au collisions at $\sqrt{s_{NN}} = 130$ GeV (from Ref. [11]). The 5% most central collisions are shown shaded.

those shown in Fig. 8.6, irrespectively of different colliding nuclei, the collision energy higher by an order of magnitude, and taking only particles emitted into a small fraction of the full space angle. The multiplicity distribution shown in Fig. 8.7 has been normalized to the total inelastic Au+Au cross section of 7.2 b. The shaded area in the plot indicates the multiplicity distribution for the 5% most central collisions which contribute 360 mb to the total cross section, and yield the largest multiplicities.

Mean multiplicities of all charged secondaries, $\langle N_{\rm ch}\rangle$, and of negatively charged secondaries, $\langle N_-\rangle$, together with corresponding values of the dispersion of multiplicity distributions, $D_{\rm ch}$, and D_-, for ^{16}O+Au "minimum bias" and "central" collisions at 200A GeV are given in Table 8.2. Figure 8.8 shows the dispersion of the multiplicity distributions of negatively charged secondary particles from collisions of ^{16}O with various nuclear targets at 60A GeV and 200A GeV plotted against the multiplicity. The dashed line has been fitted to the data on "minimum bias" collisions, showing approx-

Table 8.2 Mean multiplicity and dispersion of all charged and of negatively charged secondaries in O+Au collisions at 200A GeV (from Ref. [4])

Reaction	$\langle N_{\rm ch}\rangle$	$D_{\rm ch}$	$\langle N_-\rangle$	D_-
O+Au, min.bias	111 ± 4	90 ± 5	45.6 ± 1.6	36.0 ± 2.1
O+Au, central[a]	289 ± 3	32 ± 3	126.5 ± 1.5	16.0 ± 1.5

[a]For "central" collisions average values have been taken of two samples selected with slightly different triggers.

imate proportionality of D_- to $\langle N_-\rangle$ for this class of collisions. On the contrary, points for "central" collisions lie far from this line, in the lower right corner of the plot. Such collisions have high multiplicity and relatively small dispersion, meaning that multiplicity distributions are narrow.

A reduction of the width of multiplicity distributions in nucleus-nucleus collisions with impact parameter selection was already observed at energies of several GeV per nucleon, where these distributions were found to have Poissonian form: $D_- = \langle N_-\rangle^{1/2}$. This is shown in Fig. 8.9 for Ar+KCl collisions at 1.8A GeV [12], and for ^{12}C and ^{16}O collisions with various nuclear targets at (4.3–4.6)A GeV [13]. In the first case collisions were selected by the number of participant protons, while in the latter case "central" collisions were selected for various colliding systems.

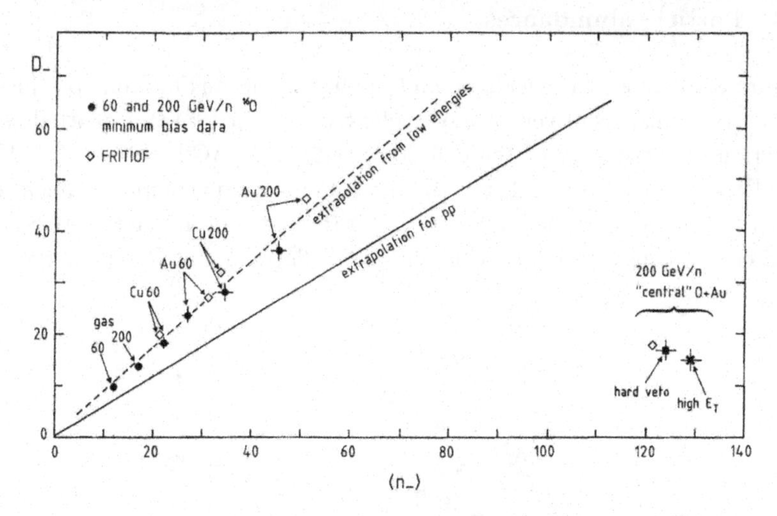

Fig. 8.8 Dispersion of multiplicity distributions of negatively charged particles plotted against the average multiplicity for "minimum bias" and "central" collisions of ^{16}O with various nuclei at 60A GeV and 200A GeV. Full line shows the dependence of D_- on $\langle N_- \rangle$ for proton–proton collisions, dashed line has been fitted to the data points for "minimum bias" nuclear collisions. Points for "central" nuclear collisions lie in the lower right corner of the plot (from Ref. [4]).

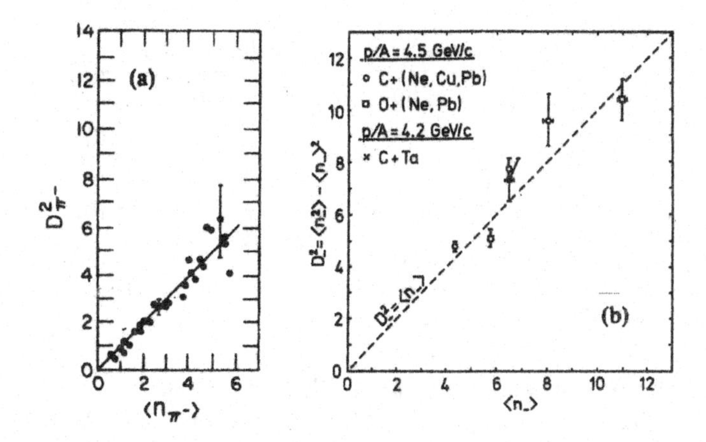

Fig. 8.9 Dispersion squared of the multiplicity distribution of negatively charged pions plotted against their average multiplicity for: (a) Ar+KCl collisions at 1.8A GeV (from Ref. [12]), and (b) for ^{12}C and ^{16}O collisions with various nuclei at (4.3–4.6)A GeV (from Ref. [13]). All events have been impact parameter selected (see text). Straight lines correspond to Poissonian distributions.

8.3 Particle abundances

Figure 8.10 shows the yields at midrapidity of various hadrons produced in central collisions of very heavy nuclei (Au+Au or Pb+Pb), plotted as a function of the energy of the collision. Data from AGS, SPS, and RHIC accelerators have been scaled to the same number of participating nucleons $N_{\text{part}} = 350$. A steep increase of all yields (except protons) in the AGS and SPS energy range is clearly seen, with some flattening at RHIC energies.

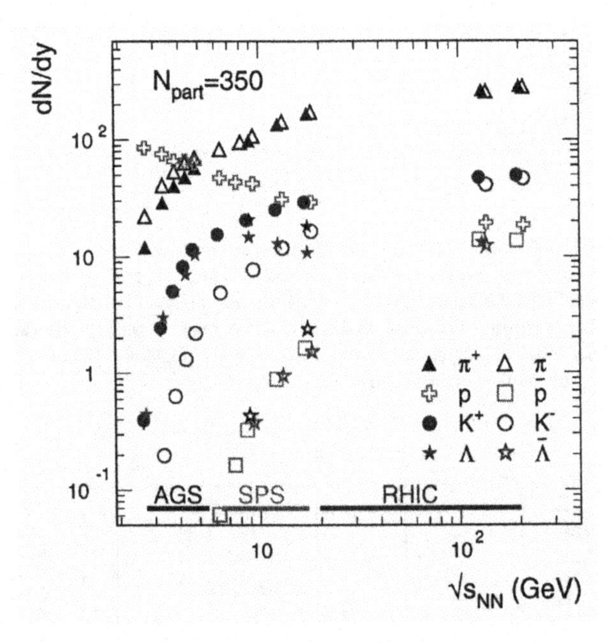

Fig. 8.10 Energy dependence of measured particle yields at midrapidity of various hadrons produced in central collisions of very heavy nuclei (Au+Au or Pb+Pb). The energy intervals covered by various accelerators (AGS, SPS, and RHIC) are marked above the horizontal scale. All data have been scaled to the same number of participating nucleons $N_{\text{part}} = 350$ (from Ref. [14]).

Particles produced in collisions of relativistic nuclei are mostly pions, charged and neutral, in approximately equal proportions. While in collisions of nuclei with $N = Z$ (light nuclei up to $^{40}_{20}$Ca), due to isospin symmetry of the initial state $\langle n_{\pi^+} \rangle = \langle n_{\pi^-} \rangle = \langle n_{\pi^0} \rangle$, some excess of negatively charged pions over positively charged ones is observed in collisions of heavier nuclei which have more neutrons than protons. This excess, relatively more important at lower collision energies, is clearly seen in Fig. 8.10. In general, the mean multiplicity of neutral pions is $\langle n_{\pi^0} \rangle = (\langle n_{\pi^+} \rangle + \langle n_{\pi^-} \rangle)/2$.

With increasing energy of a collision, heavier particles become to be produced more and more abundantly. K and $\bar{K}$ mesons appear first. Here an excess of K^+'s over K^-'s is observed due to two different production mechanisms: *the associated production* of kaons together with hyperons, $N + N \rightarrow Y + K$, which has a lower threshold and is a source of K-mesons only, while the kaon pair production, $N + N \rightarrow N + N + K + \bar{K}$, is a source of kaons and antikaons in equal proportions. Λ, Σ, and also Ξ and Ω hyperons are produced together with K-mesons in various *associated production* processes. Due to an increasing importance of the $K\bar{K}$ pair production, the excess of K^+ over K^- decreases with increasing collision energy. However, in the energy dependence of the K^+ yield a sharp maximum at $\sqrt{s_{NN}} \approx 7$ GeV (a "horn") is seen, as shown in Fig. 8.11. This anomaly, occuring at the laboratory energy of about 30A GeV, was difficult to explain, a conjecture was even made that it could indicate an energy threshold for quark-gluon plasma formation [15]. Later, however, it has been understood in the framework of the statistical-thermal model [16]. At RHIC energies the total yield of K-mesons relative to pions is about 20%. The proton yield at midrapidity decreases with increasing energy, approaching that of antiprotons at the top RHIC energy.

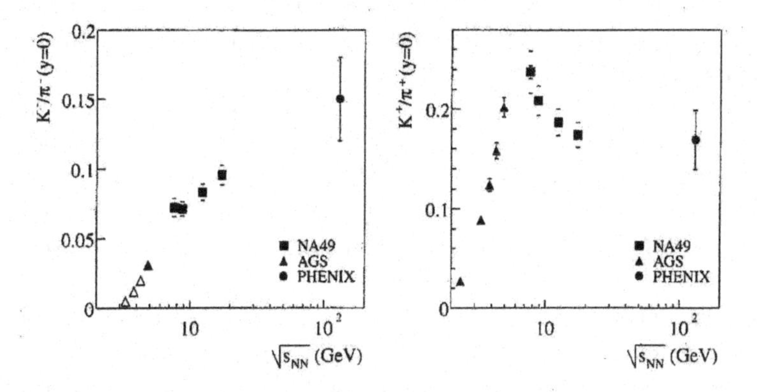

Fig. 8.11 Kaon-to-pion (K^-/π^- and K^+/π^+) yield ratios near midrapidity plotted as functions of the c.m.energy of the collision (from Ref. [17]).

Figure 8.12 shows the yield of ϕ-mesons relative to charged pions as a function of the c.m. energy of the collision. The ϕ-meson ($M = 1020$ MeV) has a small width ($\Gamma = 4$ MeV), and can be detected via its dominant decay

$\phi \to K^+K^-$. Its yield increases monotonically with increasing collision energy, similarly to that of K^--mesons, in fact the ratio ϕ/K^- remains constant within errors. At RHIC energies the yield of ϕ-mesons relative to pions is about 2%. In experiments studying the production of lepton pairs

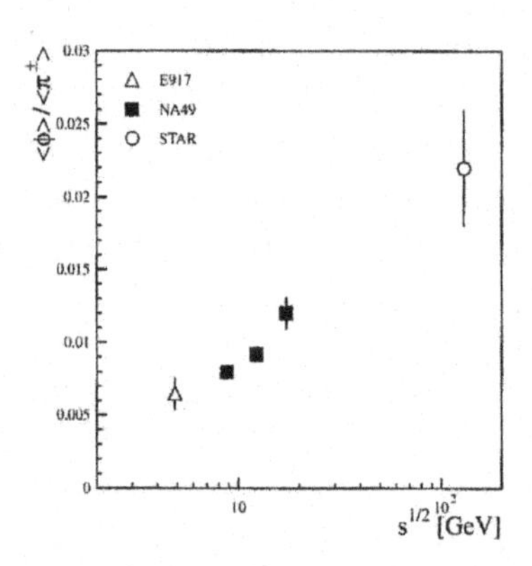

Fig. 8.12 Total $\phi/\pi^{\pm}$ yield ratio plotted as a function of the c.m.energy of the collision (from Ref. [17]).

ϕ-mesons are detected via their relatively rare ($\sim 10^{-4}$) decays $\phi \to \mu^+\mu^-$, or $\phi \to e^+e^-$. From a fit to the invariant mass distribution of lepton pairs, the ratio $\phi/(\omega + \rho)$ can be obtained[b] At SPS energies this ratio is of the order of 0.5, but, as Fig. 8.13 shows, it increases with the number of participating nucleons.

Figure 8.14 shows yields of antibaryons relative to baryons at midrapidity, plotted as a function of the c.m. energy of the collision. Yields of antibaryons rise fast with increasing collision energy, approaching at RHIC energies those of the corresponding antibaryons. This is what is expected for a baryon-free central region where antibaryons and baryons are created in pairs.

A large number of nucleons involved in a collision of nuclei, and a still larger number of secondary hadrons emerging from such collisions at relativistic energies, call for using a statistical approach to these ensembles. This approach proved extremely successful. It has turned out that the statistical-thermal model is able to fit multiplicities of various particles produced in relativistic nuclear collisions remarkably well. The analysis of relative abundances of various particles points towards a chemical equilibrium

[b] ω and ρ mesons cannot be resolved, and usually they are assumed to be produced in equal proportions.

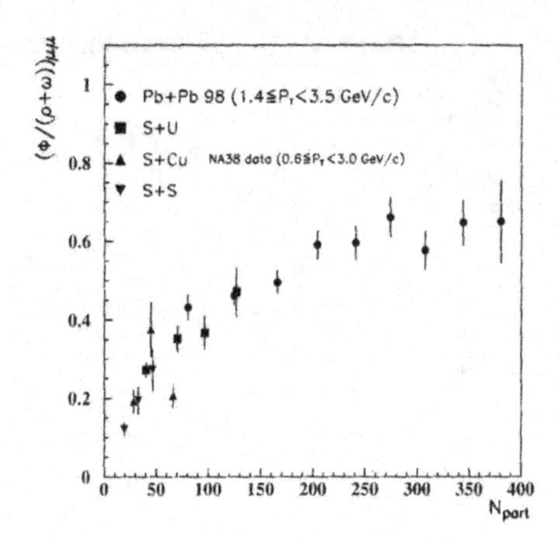

Fig. 8.13 The integrated yield ratio $\phi/(\omega + \rho)$ as a function of the number of participating nucleons (from Ref. [18]).

of the final state. This could be, but not necessarily is, a consequence of thermal equilibration of an intermediate partonic state: the quark-gluon plasma.

The statistical-thermal model has two parameters: the baryonic chemical potential μ_B, and the temperature, T. From experimental data on particle abundances the value of the baryonic chemical potential can be determined. For example, the ratio of the antiproton yield to that of protons is given by the simple expression

$$\frac{\bar{\mathrm{p}}}{\mathrm{p}} = \frac{e^{-(E+\mu_B)/T}}{e^{-(E-\mu_B)/T}} = e^{-2\mu_B/T} \tag{8.3}$$

The other parameter of the model, the temperature, can be obtained from the analysis of particle spectra, and/or from the phenomenological freeze-out condition of a fixed energy per particle, $\langle E \rangle = 1$ GeV [20]. Data sets used for this analysis are shown in Fig. 8.15, and numerical values of the statistical model parameters, together with estimates of the goodness of fit, are collected in Table 8.3.

Figure 8.16 shows these parameters in graphical form [16].

The energy dependence of the baryonic chemical potential can be parameterized as

$$\mu_B(s) \cong \frac{a}{1 + \sqrt{s}/b} \tag{8.4}$$

where $a \cong 1.27$ GeV, and $b \cong 4.3$ GeV [16]. The baryonic chemical potential decreases with increasing energy of the collision, becoming quite small at

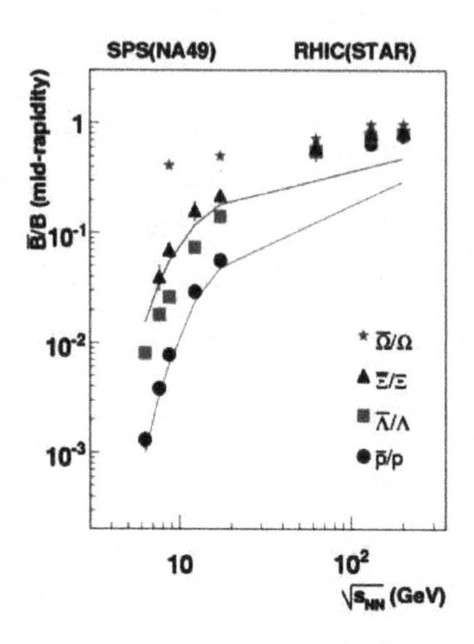

Fig. 8.14 Yield ratios at midrapidity for various antibaryons and baryons as a function of the c.m. energy per nucleon pair (from Ref. [19]).

Table 8.3 Parameters of the statistical-thermal model for central collisions of various nuclei at several energies (from Ref. [21]).

Reaction	$\sqrt{s_{NN}}$, GeV	μ_B, MeV	T, MeV	n	χ^2/n
Au+Au	4.7	540 ± 7	$125 +3/-6$		
Pb+Pb	8.7	400 ± 10	148 ± 5	11	1.1
Pb+Pb	17.3	255 ± 10	170 ± 5	24	2.0
Au+Au	130	46 ± 5	174 ± 7	13	0.8
Au+Au	200	29 ± 6	177 ± 7	5	1.1

RHIC energies, while the temperature increases, reaching a plateau value of about 170 MeV. As this value compares well with *the critical temperature* obtained from the lattice QCD calculation for the phase transition, we obtain a surprisingly consistent picture.

Besides all the particles discussed above, attempts have been made to detect various short-lived resonant states. It is well known that in elementary hadronic interactions a large fraction of secondary particles is produced via short-lived intermediate states. In nuclear collisions their detection is more difficult in view of a large combinatorial background in invariant mass distributions. Here positive results have been obtained for $\Lambda^*(1520)$ and

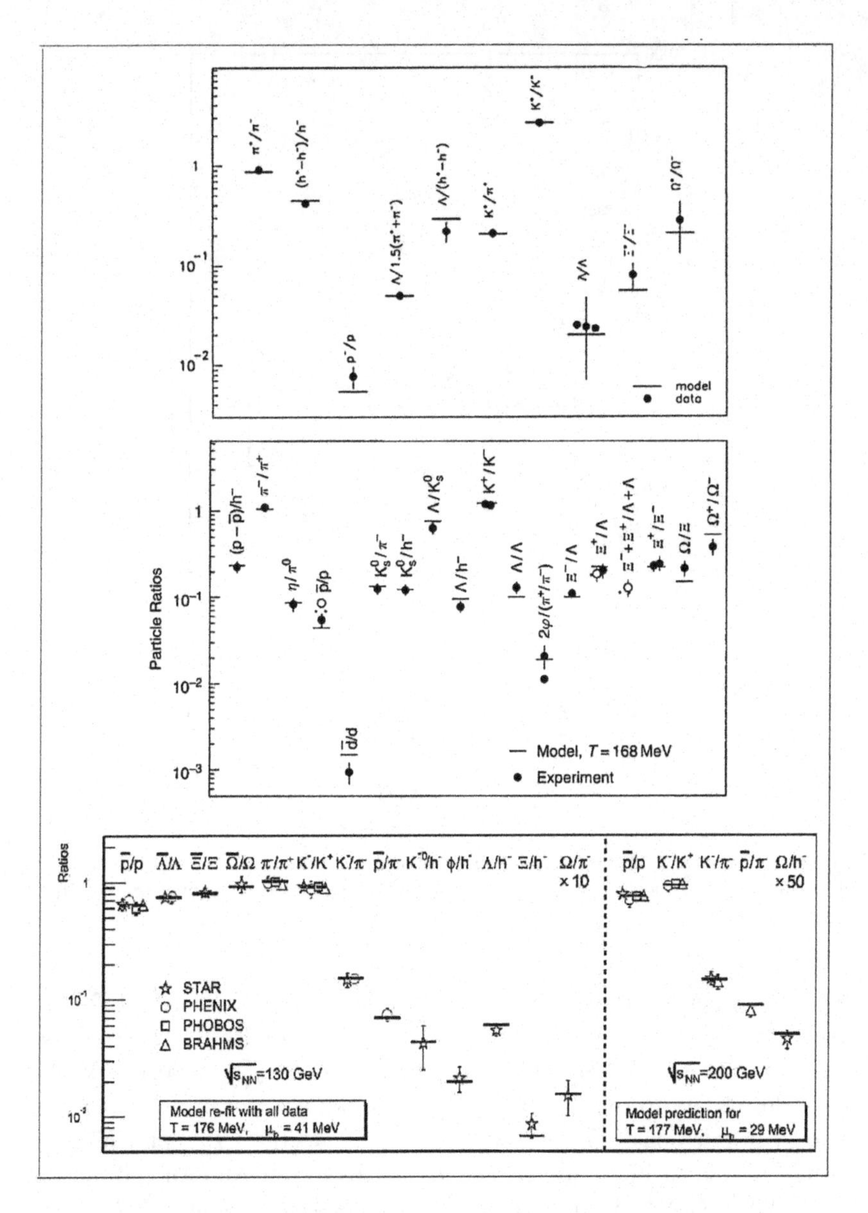

Fig. 8.15 Abundance ratios of secondary hadrons in central collisions of very heavy nuclei (open symbols), compared with the predictions of the statistical-thermal model (short horizontal lines). Upper plot: Pb+Pb at $\sqrt{s_{NN}}$ = 8.7 GeV (from Ref. [21]), middle plot: Pb+Pb at $\sqrt{s_{NN}}$ = 17.3 GeV (from Ref. [22]), lower plot: Au+Au at $\sqrt{s_{NN}}$ = 130 and 200 GeV (from Ref. [23]).

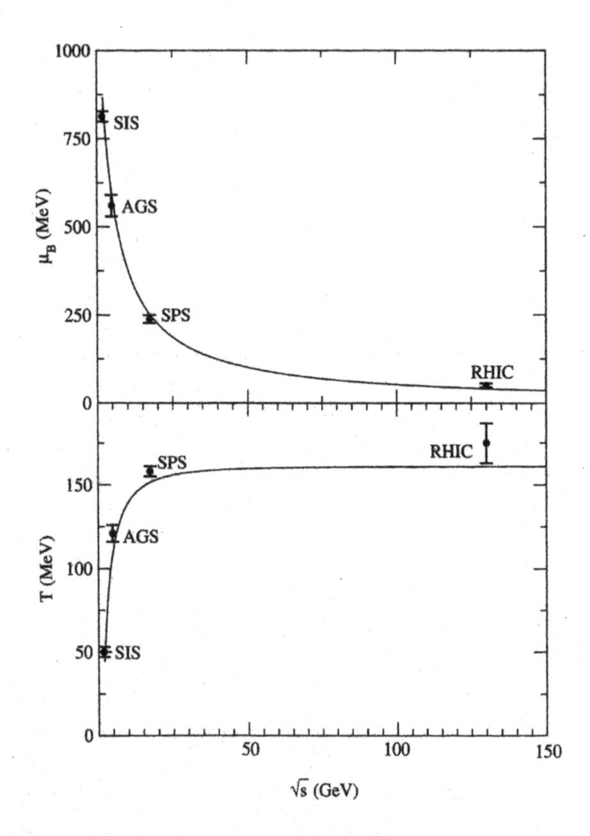

Fig. 8.16 Behaviour of the freeze-out baryon chemical potential μ_B (upper curve), and the temperature T (lower curve) as a function of the c.m.energy per nucleon pair (from Ref. [16]).

$K^*(892)$. Their widths are, respectively, 16 and 50 MeV. The $\Lambda^*(1520)$ has been detected in its decay channel $\Lambda^* \rightarrow \mathrm{p} + K^-$ in both Pb+Pb collisions at $\sqrt{s} = 17.3A$ GeV [24] and Au+Au collisions at $\sqrt{s} = 200A$ GeV [25]. Its yield relative to the Λ is $\Lambda^*/\Lambda \cong 0.02$, about a factor of two lower than in p+p interactions. The $K^*(892)$ in the charge states K^{*0} and $K^{*\pm}$ has been detected in its decay channel $K^* \rightarrow K + \pi$ in Au+Au collisions at $\sqrt{s} = 200A$ GeV [26]. Its yield relative to kaons is about $K^*/K \cong 0.2$, again about a factor of two lower than in p+p interactions. The measured yields of $\Lambda^*(1520)$ and $K^*(892)$ are also significantly lower than the thermal model predictions. A possible interpretation of this result would be the loss of the resonance signal due to scattering (elastic or inelastic) of daughter particles in the medium, occuring just after decay of a parent state. Such processes change the invariant mass of the resonance decay products, and might push it out of the resonance region, what leads to a decrease of the observed signal.

Taking this as a plausible explanation of the observed discrepancies one can conclude that multiplicities of secondary particles produced in collisions of relativistic nuclei can be understood within the framework of the statistical-thermal model.

The question of *strangeness enhancement* will be discussed in Chapter 12, and the effect of *suppression of charmonium* in Chapter 16.

References

[1] T. Alber *et al.* (NA35 Collaboration), *Eur. Phys. J. C* **2** (1998) 643.
[2] NA49 Collaboration, private communication.
[3] B. B. Back *et al.* (PHOBOS Collaboration), *Phys. Rev. C* **74** (2006) 021902, and the PHOBOS Collaboration www page.
[4] A. Bamberger *et al.* (NA35 Collaboration), *Phys. Lett. B* **205** (1988) 583.
[5] J. W. Harris *et al.*, *Phys. Lett. B* **153** (1985) 377.
[6] J. W. Harris *et al.*, *Phys. Rev. Lett.* **58** (1987) 463.
[7] L. Simič *et al.*, *Phys. Rev. D* **34** (1986) 692.
[8] B. B. Back *et al.* (PHOBOS Collaboration), *Nucl. Phys. A* **757** (2005) 28, and references therein.
[9] Z. Koba, H. B. Nielsen and P. Olesen, *Nucl. Phys. B* **40** (1972) 317; P. Slattery, *Phys. Rev. Lett.* **29** (1972) 1624.
[10] I. Lund *et al.* (WA80 Collaboration), *Z. Phys. C* **38** (1988) 51.
[11] M. Calderon de la Barca Sanchez (STAR Collaboration), *Nucl. Phys. A* **698** (2002) 503c.
[12] A. Sandoval *et al.*, *Phys. Rev. Lett.* **45** (1980) 874.
[13] J. Bartke *et al.*, *Yad. Fizika* **32** (1980) 699 [*Sov. J. Nucl. Phys.* **32** (1980) 361].
[14] A. Andronic, P. Braun-Munzinger and J. Stachel, *Nucl. Phys. A* **772** (2006) 167.
[15] M. Gaździcki and M. Gorenstein, *Acta Phys. Polon. B* **30** (1999) 2705.
[16] P. Braun-Munzinger *et al.*, *Nucl. Phys. A* **697** (2002) 902.
[17] V. Friese *et al.* (NA49 Collaboration), *J. Phys. G: Nucl. Part. Phys.* **30** (2004) S119, and references therein.
[18] D. Jouan *et al.* (NA50 Collaboration), *J. Phys. G: Nucl. Part. Phys.* **30** (2004) S277.
[19] C. Hoehne *et al.* (NA49 Collaboration), *Nucl. Phys. A* **774** (2006) 35.
[20] J. Cleymans and K. Redlich, *Phys. Rev. Lett.* **81** (1998) 5284.
[21] P. Braun-Munzinger, K. Redlich and J. Stachel, *Quark-Gluon Plasma 3*, eds. R. C. Hwa and X.-N. Wang (World Scientific, 2004), p. 491.

[22] P. Braun-Munzinger, I. Heppe and J. Stachel, *Phys. Lett. B* **465** (1999) 15.
[23] P. Braun-Munzinger *et al.*, *Phys. Lett. B* **518** (2001) 41; D. Magestro, *J. Phys. G: Nucl. Part. Phys.* **28** (2002) 1748.
[24] V. Friese *et al.* (NA49 Collaboration), *Nucl. Phys. A* **698** (2002) 487.
[25] L. Gaudichet *et al.* (STAR Collaboration), *J. Phys. G: Nucl. Part. Phys.* **30** (2004) S549.
[26] H. Zhang *et al.* (STAR Collaboration), *J. Phys. G: Nucl. Part. Phys.* **30** (2004) S577.

Chapter 9

Longitudinal Distributions of Secondary Particles

Longitudinal distributions of secondary particles from high energy reactions are usually studied in rapidity, y, or pseudorapidity, η, variables.[a] For identified particles, the "true" rapidity can be determined. This was the case for experiments at the AGS where energies of secondary particles were not very high. Figure 9.1 shows the rapidity distributions for pions, kaons, and protons from central Si+Al collisions at 14.6A GeV. These distributions are anisotropic (isotropic distributions are shown in

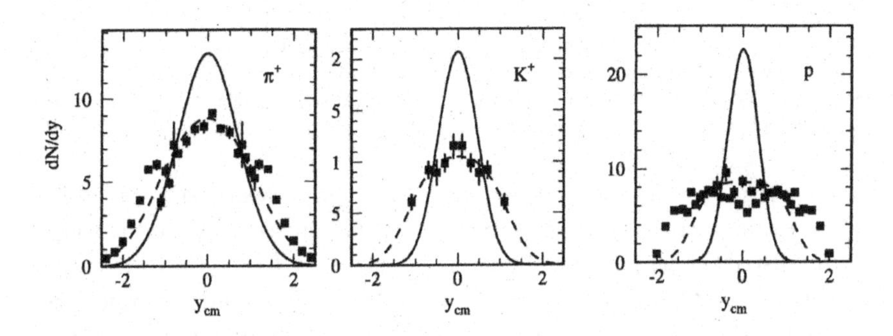

Fig. 9.1 Rapidity distributions of pions, kaons, and protons from central Si+Al collisions at 14.6A GeV (from Ref. [1]). See text for details.

Fig. 9.1 with solid lines), and for pions and kaons they can be described assuming a longitudinally expanding thermal source, the expansion velocity being $\langle\beta_L\rangle = 0.52$ (dashed lines in Fig. 9.1).[b] The width of the rapidity

[a] Sometimes also the Feynman's x-variable, defined as $x_F = p_L/p_L^{max}$, is used.

[b] The transverse spectra of pions and kaons can be described using the same model of an expanding thermal source, but the transverse expansion velocity turns out to be smaller, $\langle\beta_T\rangle = 0.33$–$0.39$.

distribution for protons is much larger, indicating an incomplete stopping of the incoming nucleons.

At higher energies the experimental conditions are different: only a fraction of secondary particles could be identified, and thus pseudorapidity is generally used. Figure 9.2 shows pseudorapidity distributions of charged hadrons (without their identification) from Au+Au collisions at two energies: $\sqrt{s} = 19.6A$ GeV and $\sqrt{s} = 200A$ GeV for different "centralities", labeled by the fraction of the total inelastic cross section. At a fixed energy

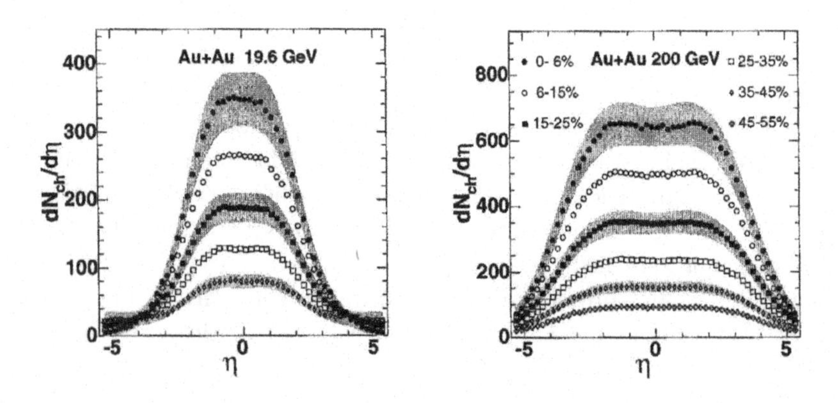

Fig. 9.2 Pseudorapidity distributions of secondary charged hadrons from Au+Au collisions at two energies, and for different centralities, labeled by the fraction of the total inelastic cross section (from Ref. [2]). Note that vertical scales differ by a factor of two, while horizontal scales are the same in both plots.

the number of secondary particles increases with increasing centrality, what is due to the increasing number of participating nucleons, while the shape of the distributions remains similar. On the other hand, with increasing energy the shape of the distribution changes: while at SPS energies distributions show a single hump centered at midrapidity, at RHIC energies they are wider, and in the central region almost flat, perhaps with a shallow minimum at midrapidity. The conjecture that at a fixed energy the production of secondary hadrons is determined by the number of participating nucleons is supported e.g. by a comparison of Au+Au and Cu+Cu collisions. In Fig. 9.3 such comparison is made, the two samples of nuclear collisions having been selected in a way to feature similar numbers of participating nucleons, $N_{\text{part}} \approx 100$. Both distributions are seen to coincide within errors.

When passing from unidentified to identified hadrons, one can *a priori* expect that longitudinal distributions of baryons should be basically differ-

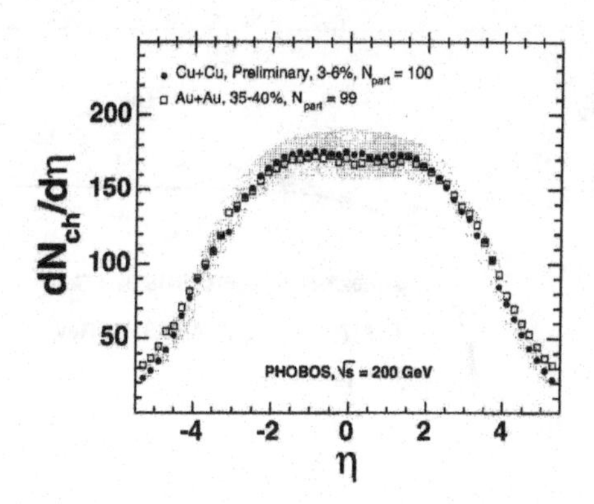

Fig. 9.3 Pseudorapidity distributions of charged hadrons from Cu+Cu and Au+Au collisions at $\sqrt{s} = 200A$ GeV. Centralities of the two samples were selected so as to yield similar N_{part}. The grey band shows the systematic uncertainty for Cu+Cu, errors for Au+Au are not shown (from Ref. [3]).

ent from those of mesons (and other newly produced particles), as nucleons from the colliding nuclei are initially situated at two opposite edges of the available longitudinal phase space, $\pm y_p$ (or y_{beam} and y_{target}), while new particles will be produced mainly in the central region. This is indeed the case, however the difference becomes really apparent only at higher energies. The initial distribution of nucleons can be described by two Gaussians centered at y_{beam} and y_{target}, their width being determined by the intranuclear Fermi motion (see Chapter 3). As a result of the interaction, these distributions suffer a shift towards midrapidity. The magnitude of this shift, $\langle \delta y \rangle$, called *the mean rapidity loss*, is the measure of *the nucleon stopping* in nuclear matter. Figure 9.4 shows this quantity as a function of the incident energy. With increasing energy $\langle \delta y \rangle$ grows almost linearly with y_p up to SPS energies where it reaches the value of about 1.7 at $\sqrt{s} = 17A$ GeV, but then this dependence flattens. At $\sqrt{s} = 200A$ GeV (RHIC) $\langle \delta y \rangle = 2.1 \pm 0.2$, and a further extrapolation yields only slightly higher values at LHC energies.

Figure 9.5 shows rapidity distributions of "net protons", i.e. "protons minus antiprotons", for Au+Au collisions at several incident energies.[c] With increasing energy, a relatively narrow distribution becomes wider, and at RHIC energies a central region with a low net baryon number becomes

[c] At lower energies nucleons are the only carriers of the baryonic number, at higher energies a contribution from hyperons (i.e. "hyperons minus antihyperons") should also be taken into account.

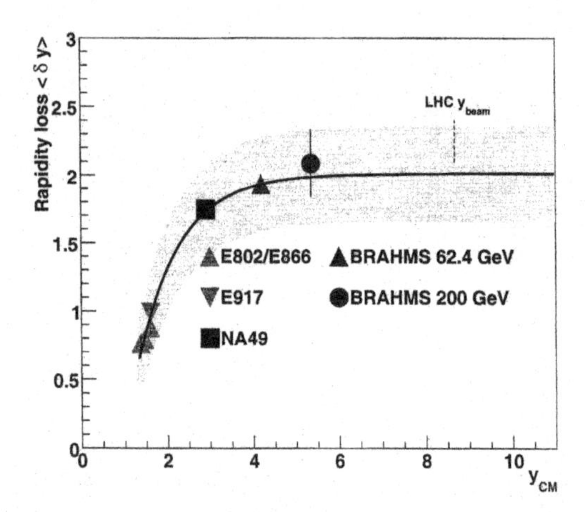

Fig. 9.4 The mean rapidity loss as a function of the incident energy (from Ref. [4]). Beam rapidity in the c.m. frame, $y_{\rm CM}$, is used as a measure of the incident energy.

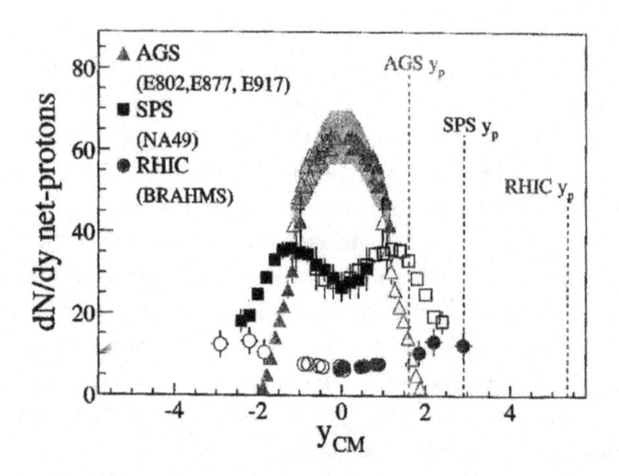

Fig. 9.5 The net proton rapidity distributions measured at AGS ($\sqrt{s} = 5.5A$ GeV), SPS ($\sqrt{s} = 17A$ GeV), and RHIC ($\sqrt{s} = 200A$ GeV) for 5% most central collisions of very heavy ions (Au+Au or Pb+Pb). For RHIC data the closed symbols indicate measured points and open symbols are symmetrized (reflected), while the opposite is true for AGS and SPS data (from Ref. [5]).

pronounced. At LHC energies a baryon-free ($B \approx 0$) region, about ten units of rapidity long, can be expected.

Figure 9.6 shows the rapidity distributions of various secondary particles, and of their antiparticles, from central Au+Au collisions at $\sqrt{s} =$

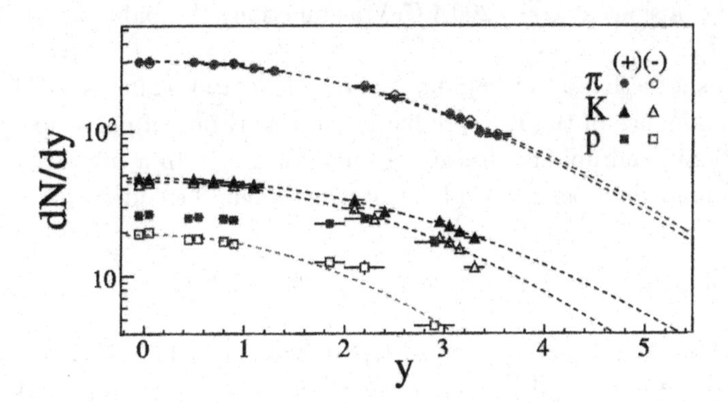

Fig. 9.6 Multiplicity densities of various secondary particles from Au+Au central collisions at $\sqrt{s} = 200A$ GeV as functions of rapidity. The lines show Gaussian fits to experimental data. No such fit was attempted for protons (from Ref. [6]).

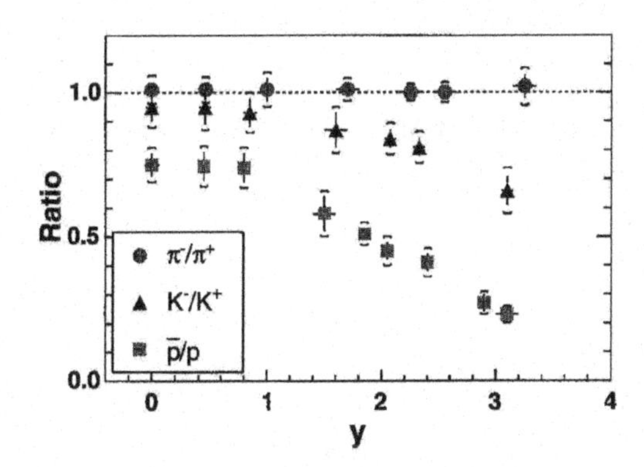

Fig. 9.7 Ratios of pions, kaons, and antiprotons to their antiparticles in Au+Au collisions at $\sqrt{s} = 200A$ GeV plotted as functions of rapidity (from Ref. [7]).

200A GeV. One can see that while longitudinal distributions of π^+ and π^- coincide, those of K^+ and K^- are different, with an excess of K^+'s at large rapidities. This effect can be understood in terms of two production mechanisms of K^+'s: the $K\bar{K}$ pair production occuring in the central region, and the associated KY production of kaons together with hyperons occuring in the baryon-rich region, i.e. at large rapidities. A difference between distributions of protons and antiprotons is also clearly seen, antiprotons being confined to the central region. The ratios π^-/π^+, K^-/K^+, and $\bar{\text{p}}/\text{p}$ in

Au+Au collisions at $\sqrt{s} = 200A$ GeV as functions of rapidity are displayed in Fig. 9.7.

Let us note that at midrapidity the K^-/K^+ ratio is almost one, indicating a dominance of the $K\bar{K}$ production in this region of phase space. The $\bar{\mathrm{p}}/\mathrm{p}$ ratio at midrapidity is also not far from unity. In a fully baryon-free region numbers of baryons and antibaryons should be equal.

References

[1] P. Braun-Munzinger *et al.*, *Phys. Lett. B* **344** (1995) 43.
[2] B. B. Back *et al.* (PHOBOS Collaboration), *Phys. Rev. Lett.* **91** (2003) 052303.
[3] G. Roland (PHOBOS Collaboration), *Nucl. Phys. A* **774** (2006) 113.
[4] I. G. Bearden (BRAHMS Collaboration), *J. Phys. G: Nucl. Part. Phys.* **34** (2007) S207.
[5] I. G. Bearden *et al.* (BRAHMS Collaboration), *Phys. Rev. Lett.* **93** (2004) 102301; P. Christensen, Ph.D. Thesis, University of Copenhagen (June 2003).
[6] D. Ouerdane, Ph.D. Thesis, University of Copenhagen (August 2003); I. G. Bearden *et al.* (BRAHMS Collaboration), nucl-ex/0403050.
[7] I. G. Bearden *et al.* (BRAHMS Collaboration), *Phys. Rev. Lett.* **90** (2003) 102301.

Chapter 10

Transverse Spectra of Secondary Particles

Spectra of particles produced in collisions of relativistic nuclei are usually displayed in "transverse" variables: transverse momentum $p_T = p \sin\theta$ where θ is the emission angle with respect to the collision axis, or "transverse mass" $m_T = \sqrt{p_T^2 + m_0^2}$ where m_0 is the particle rest mass. Both these variables are Lorentz-invariant. The use of m_T is suggested by thermal emission models which predict a simple exponential shape of the spectra in m_T. Also, some experiments at colliders can measure identified particle spectra only in a narrow angular interval around the central value of rapidity ($|\eta| \leq 1.0$ in STAR, $|\eta| \leq 0.9$ in ALICE), and thus the measured energy spectra are, in fact, the m_T spectra, as $E = (p^2 + m_0^2)^{1/2} = (p_L^2 + p_T^2 + m_0^2)^{1/2} \approx m_T$ for small values of the longitudinal momentum component p_L.

For comparing spectra of particles with different masses, it is convenient to replace m_T by $(m_T - m_0)$. When plotted in this variable, spectra of different particle species all begin at zero on the horizontal scale, and their slopes and detailed shapes can be well compared.

Spectra of produced particles are usually fitted with the formula[a]

$$\frac{1}{m_T}\frac{d\sigma}{dm_T} = C \exp\left[-(m_T - m_0)/T\right] \tag{10.1}$$

where the inverse slope parameter T is commonly called "temperature" of the emitting source.

In early experiments with relativistic nuclei, performed at lower energies and with visual detectors, spectra of negatively charged particles were mainly investigated as representing the "produced mesons". With kaons constituting a few-percent admixture, and with a negligible amount of an-

[a]Some authors prefer to use the formula with $1/m_T^{3/2}$ instead of $1/m_T$, what results in slightly lower values of T obtained from the fitting procedure.

tiprotons, such spectra represent quite well those of produced pions. The close-to-exponential shape of secondary particles spectra was noticed already in those experiments. With the development of more sophisticated detection techniques, spectra of various types of particles can be determined separately. As an example, Fig. 10.1 shows the spectra of six types of particles from the Pb+Pb collisions at 158A GeV. The main features of these

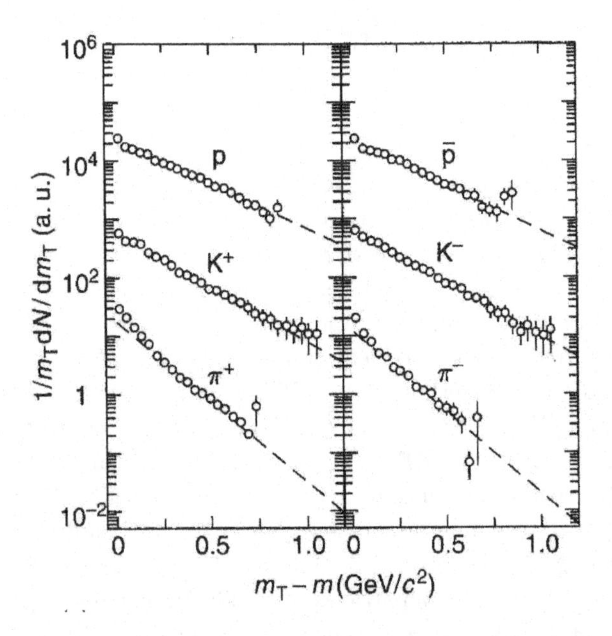

Fig. 10.1 Transverse mass spectra of pions, kaons, protons and their antiparticles from Pb+Pb collisions at 158A GeV. Spectra of K^+/K^- and p/$\bar{\text{p}}$ have been shifted upwards for clarity. Dashed lines are the exponential fits (from Ref. [1]).

spectra are: close-to-exponential shape, similar slopes for particles and antiparticles, and increase of the inverse slope parameter T with increasing particle mass.

The dependence of the inverse slope parameter T on the particle mass is shown in Figs. 10.2 and 10.3. Figure 10.2 shows that this dependence, absent in p+p collisions, becomes stronger with increasing masses of the colliding nuclei. Figure 10.3 shows the dependence of T on m for central Pb+Pb collisions at 158A GeV. This dependence is suggestive of a transversely expanding source. An expansion with a velocity β_T would cause a linear increase of the apparent "temperature", T_{app}, with increasing particle

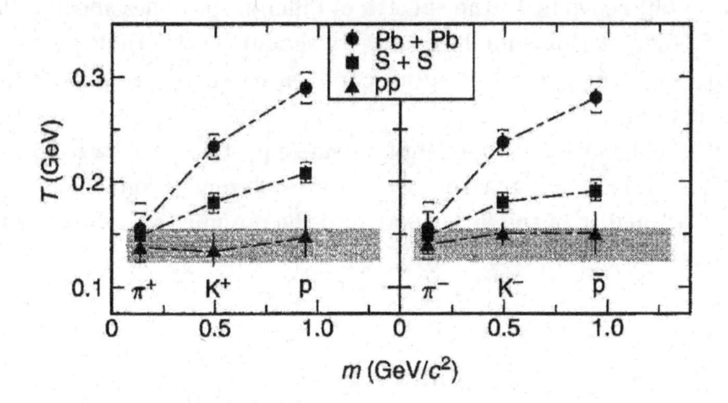

Fig. 10.2 Inverse slope parameter T for pions, kaons, protons, and their antiparticles in p+p, S+S, and Pb+Pb collisions at 158A GeV (from Ref. [1]).

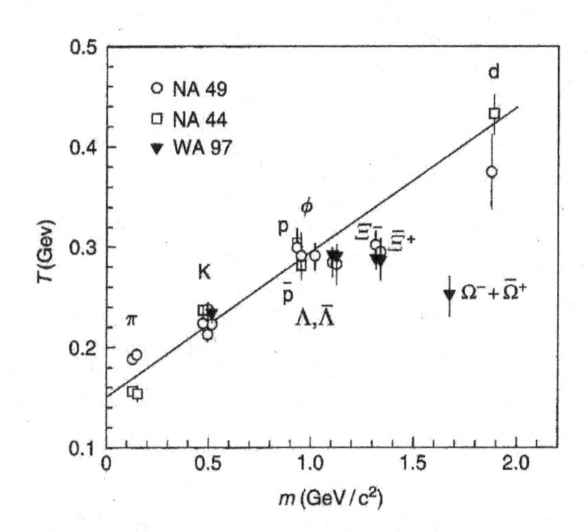

Fig. 10.3 Dependence of the inverse slope parameter T on particle mass for central Pb+Pb collisions at 158A GeV (from Ref. [2]).

mass, what is indeed observed in Fig. 10.3. The relevant phenomenological model is a "blast wave" [3], with the expansion velocity assumed to increase with the radius r: $\beta_T = \beta_s\, r/R_G$ where β_s is the flow velocity at the surface, and R_G is the outer radius of the expanding fireball. A fit to the experimental spectra yields the values of T and of the average transverse flow velocity $\langle\beta_T\rangle = \frac{2}{3}\beta_s$ [4]. Common values of T and of $\langle\beta_T\rangle$ obtained from a simultaneous fit to the spectra of different particles speak in favour of the validity of this simple model. It should be mentioned that slight deviations of some spectra from an exponential shape are also accounted for by the model.

Figure 10.4 shows, as an example, strange particle spectra from Pb+Pb collisions at 158A GeV, together with curves obtained from the blast-wave model. The quality of the fit is good, and the obtained values of the model parameters are $T = (144 \pm 7)$ MeV, $\langle\beta_T\rangle = 0.38 \pm 0.01$.

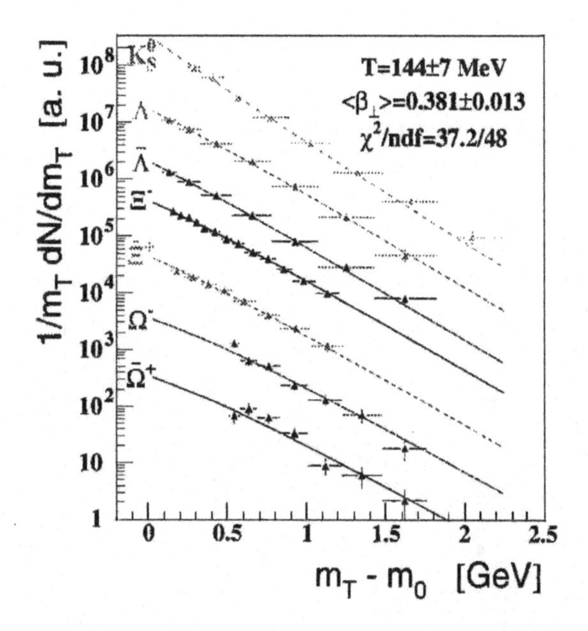

Fig. 10.4 Transverse mass spectra of strange particles for the 53% most central Pb+Pb collisions at 158A GeV, together with blast-wave fits (from Ref. [5]).

Figure 10.5 shows the dependence of T and β_T on the number of nucleons, N_{part}, participating in the interaction, for Au+Au collisions at $\sqrt{s} = 200A$ GeV. The values of T lie in the range of 120–140 MeV, similar as at SPS energies, what is consistent with the interpretation of T as the "universal" thermal freeze-out temperature. With increasing centrality of a collision T decreases slightly, while the radial flow velocity β_T increases and reaches the values 0.70c–0.75c, significantly larger than at SPS ener-

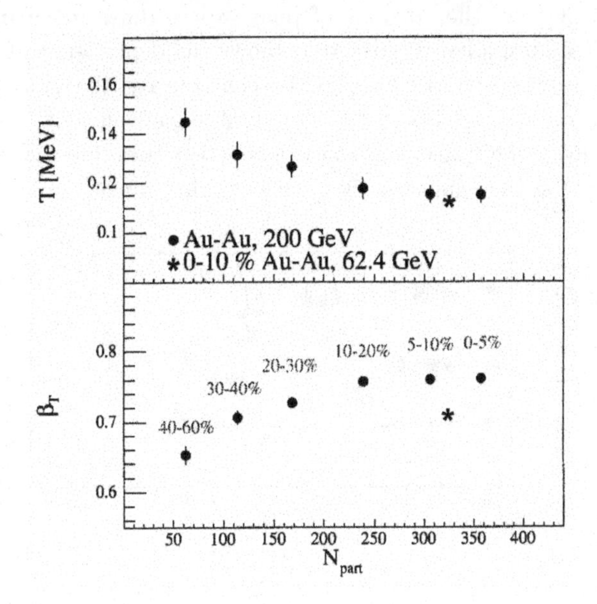

Fig. 10.5 Blast-wave model fit parameters T and β_T as functions of the number of nucleons participating in a Au+Au collision at $\sqrt{s} = 200A$ GeV. A single point for central Au+Au collisions at $\sqrt{s} = 62.4A$ GeV is also shown (from Ref. [6]).

gies. This last feature can be understood as resulting from a higher initial state density created in nuclear collisions at higher energies, which leads to a higher pressure gradient.

At SPS energies the transverse momentum spectra have been measured up to $p_T = 2$–3 GeV/c, at RHIC this limit was pushed up to $p_T \approx 10$ GeV/c for charged hadrons (STAR), and up to $p_T \approx 20$ GeV/c for π^0's (PHENIX). It is a common conjecture that high p_T particles arise not from "thermal emission", but from another mechanism: "hard collisions" and/or "parton cascade". The contribution of "hard collisions", small at SPS energies, should increase with increasing incident energy, and at RHIC energies is estimated to be at the 10% level [7]. As "hard collisions" produce more secondary particles, their relative contribution to particle spectra would be even more substantial. Nevertheless, the p_T spectra at RHIC continue to show an exponential shape also for higher p_T. Some change in this behaviour might occur at the LHC, where "hard processes" would probably play a dominant role.

It is, of course, interesting to check whether transverse spectra of secondary particles in other regions of phase space differ much from those measured at midrapidity. Figure 10.6 shows the dependence of the mean value of the transverse mass, $\langle m_T \rangle$, which characterizes the width of the m_T distribution, on the rapidity, y. A very weak dependence of $\langle m_T \rangle$ on y is seen, with only a slight decrease towards $y = 3$–4, meaning that transverse spectra do not change appreciably in this rapidity range.

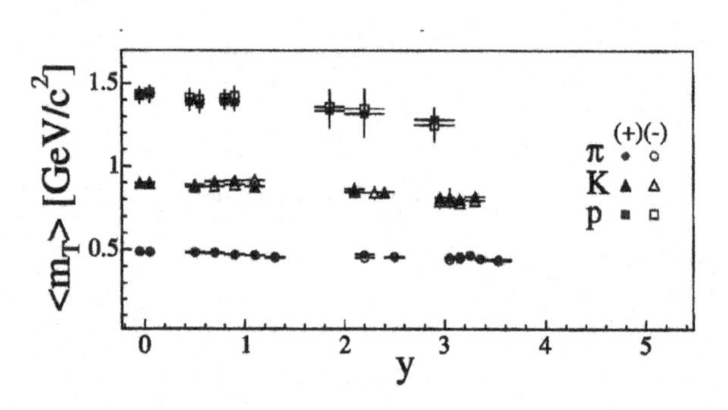

Fig. 10.6 Mean value of the transverse mass as a function of rapidity for pions, kaons, protons and their antiparticles from central Au+Au collisions at $\sqrt{s} = 200A$ GeV (from Ref. [8]).

References

[1] I. G. Bearden *et al.* (NA44 Collaboration), *Phys. Rev. Lett.* **78** (1997) 2080.
[2] F. Antinori *et al.* (WA97 Collaboration), *Eur. Phys. J. C* **14** (2000) 633.
[3] E. Schnedermann *et al.*, *Phys. Rev. C* **48** (1993) 2462; *ibid.* **50** (1994) 1675.
[4] F. Antinori *et al.* (NA57 Collaboration), *J. Phys. G: Nucl. Part. Phys.* **30** (2004) 823.
[5] A. Dainese (NA57 Collaboration), *Nucl. Phys. A* **774** (2006) 51.
[6] P. Staszel (BRAHMS Collaboration), *Nucl. Phys. A* **774** (2006) 77.
[7] D. Kharzeev and M. Nardi, *Phys. Lett. B* **507** (2001) 121.
[8] D. Ouerdane, Ph.D. Thesis, University of Copenhagen (August 2003); I. G. Bearden *et al.* (BRAHMS Collaboration), nucl-ex/0403050.

Chapter 11

Electromagnetic Effects on Charged Meson Spectra

Influence of the electric charge of the target nucleus on the energy spectra of secondary particles emitted in nuclear reactions has been known since more than 50 years. In nuclear emulsions exposed to cosmic rays, it was found that among low energy ($E_{\text{kin}} \cong$ few MeV) pions there is more π^- than π^+ [1]. This charge asymmetry increases with the increasing atomic number of the target nucleus, similarly as in the earlier exposure of emulsions to 390 MeV α-particles from the cyclotron. It has been explained by the Coulomb effect: positively charged mesons leaving the nucleus acquire an additional energy by Coulomb repulsion, while for negatively charged ones the opposite situation occurs. This conjecture was later confirmed by a similar study in emulsions exposed to 9 GeV protons from the synchrophasotron [2]. In this experiment it was found that the mean kinetic energy of positively charged slow pions is larger than that of negatively charged ones, and the π^-/π^+ ratio decreases with increasing pion energy, the charge asymmetry disappearing at $E_{\text{kin}} \sim 20$ MeV.

In experiments with relativistic heavy ions, aimed mainly at the quest for quark-gluon plasma, such effects have been considered to be of marginal importance. It turned out, however, that they deserve a certain attention. In collisions of heavy nuclei a large amount of electric charge is present. In a central collision of two nuclei of atomic number Z, the total charge of the created fireball will be $\sim 2Ze$, or about $160e$ for an Au+Au or Pb+Pb collision. This charge should have an effect on spectra of secondary charged particles. Due to the long-range nature of the electromagnetic interaction, charged particles will feel it even after the freeze-out and their spectra will be distorted.

In a peripheral collision, the charge of the central fireball will be smaller, but an appreciable charge will be carried by the spectator parts of the colliding nuclei. This charge should also have an effect on spectra of charged secondary particles, specifically in some regions of phase space.

The ratio of the spectra of negatively and positively charged pions emitted near midrapidity has been studied in central Au+Au collisions at the AGS, and in central Pb+Pb collisions at the SPS [3]. Figure 11.1 shows this ratio as a function of the pion kinetic energy[a] in central Au+Au collisions at 11.6A GeV [4]. For slow pions the π^-/π^+ ratio is significantly

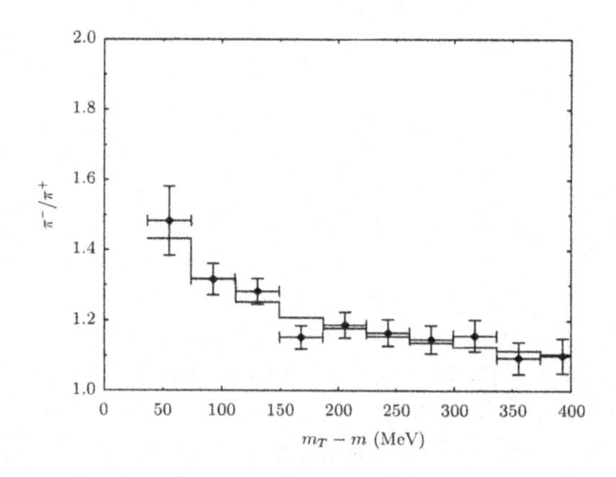

Fig. 11.1 The π^-/π^+ ratio as a function of transverse mass in central Au+Au collisions at 11.6A GeV. The histogram has been obtained assuming the model of a radially expanding fireball of 10 fm radius at pion freeze-out (from Ref. [4]).

higher than for more energetic ones. This effect can be explained by electromagnetic interaction between pions and the central fireball, and allows to determine its radius at the pion freeze-out to be about 10 fm (histogram in Fig. 11.1).[b] This is about twice the radius of the gold nucleus and thus the initial fireball must have expanded to reach this size. The expansion velocity can be estimated from the analysis of particle transverse spectra (see Chapter 10), giving a consistent picture of a radially expanding hot system.

Figure 11.2 shows the ratio of the kinetic energy spectra of negatively and positively charged pions emitted near midrapidity in nuclear collisions at the SPS. The three presented samples are: central S+S and S+Pb collisions at 200A GeV, and central Pb+Pb collisions at 158A GeV. For Pb+Pb

[a] At midrapidity $p_T = p$, $m_T = E$, and $m_T - m = E_{\rm kin}$.

[b] It should be mentioned that the overall excess of negative pions over positive ones results from isospin effects due to neutron-over-proton excess in heavy nuclei. This effect is of the order of $[(N+Z)/2Z - 1] \approx 25\%$ for Au+Au or Pb+Pb collisions.

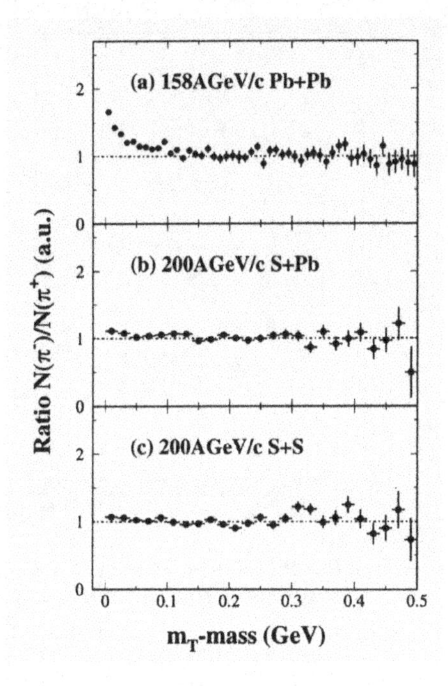

Fig. 11.2 The π^-/π^+ ratio for: (a) 158A GeV Pb+Pb, (b) 200A GeV S+Pb, and (c) 200A GeV S+S central collisions, plotted as a function of the pion transverse mass. The data have been arbitrarily normalized to unity at high m_T (from Ref. [5]).

collisions a pronounced enhancement of this ratio appears at low values of $m_T - m$, while the ratios for lighter systems are almost flat. Again, the effect observed for Pb+Pb can be explained by the Coulomb interaction [5].

In Au+Au collisions at the AGS (10.8A GeV) another effect was also observed [6]. A depletion in the π^+/π^- ratio (or enhancement in the π^-/π^+ ratio) at small m_T was found for rapidities close to the beam rapidity ($y = 3.14$). It has been attributed to the Coulomb effect, but in this case it would be the interaction with the spectator parts of the colliding nuclei.

This interesting observation was confirmed for Pb+Pb collisions at 158A GeV [7]. Here a strong enhancement in the π^-/π^+ ratio for $p_T \approx 0$ was observed at $y \approx y_{\text{beam}} = 5.7$. The effect is substantial: the π^-/π^+ ratio reaches the value of ~ 7 for the most peripheral collisions, and smoothly decreases with the increasing centrality of the collision. This is shown in Fig. 11.3 for two values of rapidity. One can see that at $y = 6.4$ the effect is much weaker than at $y = 5.7$.

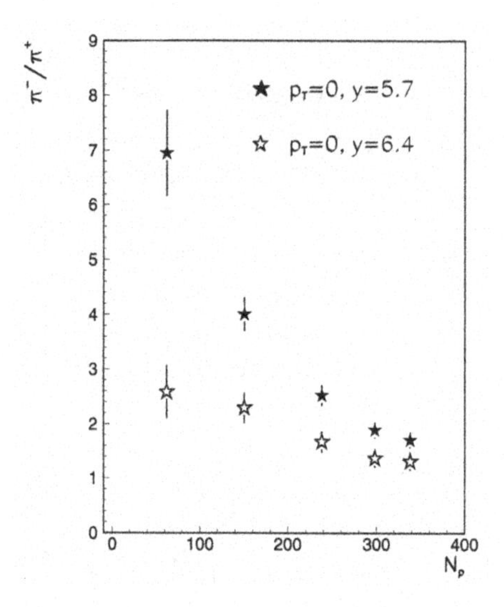

Fig. 11.3 The π^-/π^+ ratio in Pb+Pb collisions at 158A GeV for $p_T \approx 0$ as a function of the mean number of participating nucleons at rapidities $y = 5.7$ and $y = 6.4$ (from Ref. [7]).

This rapidity and centrality dependence of the π^-/π^+ ratio further supports the hypothesis of the Coulomb interaction of charged pions with the projectile spectators. In the most peripheral collisions the number of spectator nucleons, about one-half of them being protons, is the largest, and thus their effect on pion spectra should be the strongest. For the most central collisions the opposite situation takes place.

A systematic study of the Coulomb effect on charged pion spectra in Pb+Pb collisions at 158A GeV has been recently undertaken in Ref. [8]. A two-dimensional study (in x_F and p_T) gives a more detailed insight into this phenomenon. Figure 11.4 shows the results, this time for the π^+/π^- ratio. A deep minimum in this ratio appears at $x_F \approx 0.15$, and at small p_T. Theoretical calculation of electromagnetic interaction between the produced charged pions and the spectator parts of incident nuclei in a peripheral Pb+Pb collision, presented in Refs.[8, 9], yields a very similar pattern. Moreover, it shows that the effect of the spectator matter electric charge on the pion spectra has some sensitivity to initial conditions of pion production, such as the source size and the emission time. Thus, investigation of the Coulomb effect on charged particle spectra provides a different, and hopefully complementary, approach with respect to correlation stud-

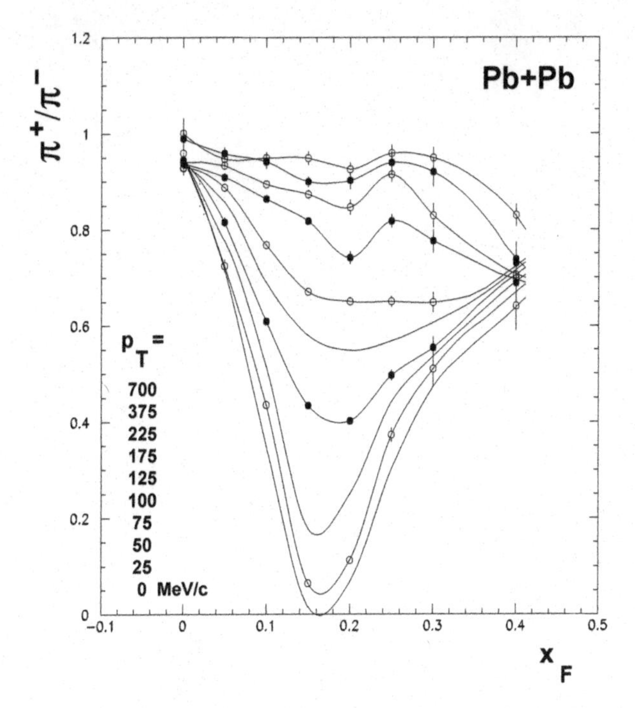

Fig. 11.4 The π^+/π^- ratio as a function of x_F for various values of p_T in the range $0 \leq p_T \leq 700$ MeV/c (from Ref. [10]).

ies which since a long time have been used to obtain information on the space-time particle distribution at freeze-out (see Chapter 14).

References

[1] H. Yagoda, *Phys. Rev.* **85** (1952) 891.
[2] E. M. Friedländer, *Phys. Lett.* **2** (1962) 38.
[3] A. Ayala *et al.*, *Phys. Rev. C* **59** (1999) 3324.
[4] L. Ahle *et al.* (E866 Collaboration), *Phys. Rev. C* **57** (1998) R466.
[5] N. Xu *et al.* (NA44 Collaboration), *Nucl. Phys. A* **610** (1996) 175c.
[6] R. Lacasse *et al.* (E877 Collaboration), *Nucl. Phys. A* **610** (1996) 153c.
[7] G. Ambrosini *et al.* (NA52 Collaboration), *New J. Phys.* **1** (1999) 23.
[8] A. Rybicki, *Int. J. Mod. Phys. A* **22** (2007) 659.
[9] A. Rybicki and A. Szczurek, *Phys. Rev. C* **75** (2007) 054903.
[10] D. Varga *et al.* (NA49 Collaboration), CERN-SPSC-2007-031 (preliminary data).

Chapter 12

Production of Strangeness and Heavy Flavours

12.1 Strangeness

Strangeness enhancement in collisions of high energy nuclei relative to elementary reactions was proposed already in the 1980s as a signature of a phase transition to quark-gluon plasma which was expected to occur in such collisions [1]. In view of these predictions, the production of strange particles has been intensively investigated by several experimental groups working at AGS, SPS, and RHIC accelerators.

As a rule, strange particles are being identified by their decays. For charged kaons, K^+ and K^-, this is difficult because of their long lifetime ($c\tau = 3.71$ m), and they are usually identified by the combined momentum and time-of-flight, or dE/dx, measurements. Nevertheless, in Ref. [2] it has been shown that decays in flight of charged kaons, appearing as "kinks" on tracks, can also be used for their indentification. This is based on the fact that in the dominant kaon decays $K \to \mu\nu$ and $K \to \pi\pi$, transverse momenta of the decay products can assume values up to, correspondingly, 236 and 205 MeV/c, substantially higher than the maximum value of 30 MeV/c in decays of pions $\pi \to \mu\nu$ which constitute the main background. Thus all "kinks" with transverse momentum exceeding 30 MeV/c can only be decays of kaons, and the loss due to this cutoff can be easily corrected for.

The main objects in investigations of strangeness production are neutral kaons K_s^0, and Λ, Σ, Ξ, Ω hyperons, and their antiparticles. They all have similar characteristic decay lengths, $c\tau \sim$ few cm, what makes their detection (except for Σ hyperons) relatively easy. The commonly studied decay channels are listed in Table 12.1[a] Finding a secondary (decay) vertex, and

[a] We quote rounded values for clarity, for more accurate values and their errors see

Table 12.1 Commonly studied strange particles.

Particle	Mass, MeV	$c\tau$, cm	Decay channels	Fraction, %
K_s^0	497.6	2.68	$\pi^+\pi^-$	69.2
$\Lambda^0, \bar{\Lambda}^0$	1115.7	7.89	$p\pi^-$, $\bar{p}\pi^+$	63.9
$\Xi^-, \bar{\Xi}^+$	1321.3	4.91	$\Lambda^0\pi^-, \bar{\Lambda}^0\pi^+$	99.9
$\Omega^-, \bar{\Omega}^+$	1672.4	2.46	$\Lambda^0 K^-, \bar{\Lambda}^0 K^+$	67.8

performing a kinematic fitting to it of tracks of the decay products, makes the identification of the parent particle possible. In a dense track environment which occurs in relativistic nuclear collisions, finding a secondary vertex is, however, not feasible. Then, a "statistical" separation of various types of strange particles can be attempted. Such method was invented by Podolanski and Armenteros in 1954, in early studies of strange particle decays in a cloud chamber exposed to cosmic rays [4]. It uses the α and p_T variables, α being defined as $\alpha = (p_+ - p_-)/(p_+ + p_-)$, with p_+ and p_- being, correspondingly, longitudinal momentum components of positively and negatively charged decay products, and p_T being the transverse momentum in the decay. Figure 12.1 shows the two-body decay kinematics in the laboratory system (LS) and in the centre-of-mass system (CMS), and defines the p_+ and p_- variables. Different types of strange particles: K_s^0, Λ,

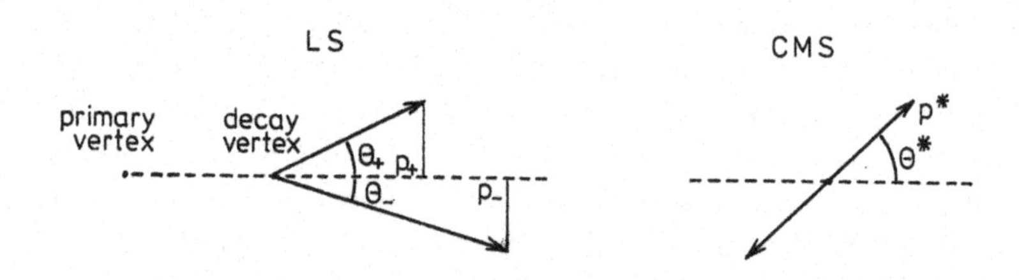

Fig. 12.1 Decay kinematics in LS and CMS.

and $\bar{\Lambda}$, when plotted in the $\alpha - p_T$ plane, fall on elliptical contours shown in Fig. 12.2.

The Armenteros–Podolanski method was successfully used also in accelerator experiments. We quote data of the NA49 experiment at the CERN

Ref. [3]. Σ hyperons are not included in the Table, as identification of decays of $\Sigma^\pm$ hyperons is in practice very difficult ("kinks" on tracks, close to the primary vertex), and Σ^0 hyperons are, due to their very short lifetime, mixed with Λ hyperons.

SPS. All random pairs of positively and negatively charged particles emerging from Pb+Pb collisions at 158A GeV have been taken for the plot shown in Fig. 12.3. Accumulation of points along elliptical contours shown in Fig. 12.2 appears clearly, allowing to estimate production rates of three types of strange particles: K^0_s, Λ, and $\bar{\Lambda}$. One should, however, keep in mind that $\Lambda(\bar{\Lambda})$ hyperons are not only the ones directly produced, but contain also the decay products of $\Sigma^0(\bar{\Sigma}^0)$ hyperons which cannot be separated because of their very short lifetime.

Identification of $\Xi(\bar{\Xi})$ and $\Omega(\bar{\Omega})$ hyperons is more difficult, as it requires the reconstruction of a (secondary) $\Lambda(\bar{\Lambda})$ as a first step, and then, if its line of flight "misses" the primary vertex, associating it, correspondingly, with an appropriate $\pi^-(\pi^+)$ or $K^-(K^+)$ track.

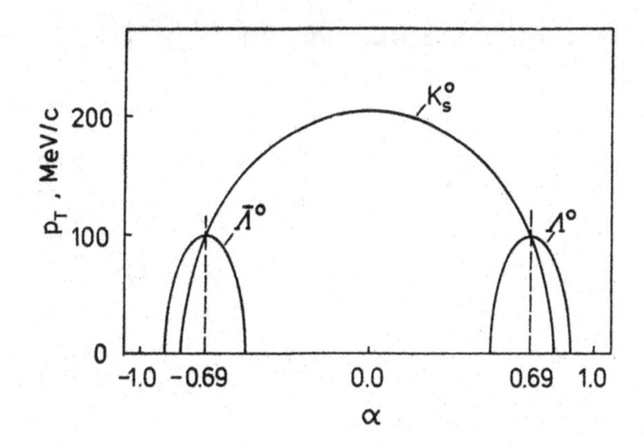

Fig. 12.2 The Armenteros–Podolanski plot.

Various measures of the strangeness content among the produced particles can be used. The simplest one is the K/π ratio, a better one is the λ_s parameter, introduced in [6], and based on quark counting

$$\lambda_s = \frac{2\langle s\bar{s}\rangle}{\langle u\bar{u}\rangle + \langle d\bar{d}\rangle} \tag{12.1}$$

This parameter measures the relative strangeness content in a given reaction. In order to correct for unobserved decay channels, the strangeness yield, $\langle s\bar{s}\rangle$, appearing in the numerator is evaluated as

$$\langle s\bar{s}\rangle = 1.6\langle\Lambda\rangle + 1.6\langle\bar{\Lambda}\rangle + 4\langle K^0_s\rangle \tag{12.2}$$

where $\langle X\rangle$ denotes the experimentally observed mean number of decays of strange particles of type X.

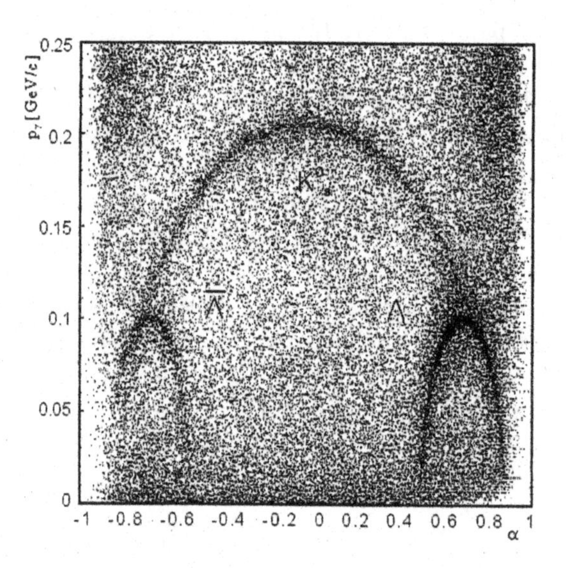

Fig. 12.3 The Armenteros–Podolanski plot for real data from the NA49 experiment (from Ref. [5]).

In p+p and p+A reactions the yield of strange quark pairs is about 22% of that of light quark pairs, and nearly independent of $\sqrt{s}$. In heavy ion collisions a much higher strangeness yield is observed. Enhancement in the K/π ratio was seen for the first time by the E802 experiment at the AGS, the E859 and NA35 reported confirmation of the factor of two enhancement in collisions of heavy nuclei over p+p and p+A. A very strong effect has been observed for hyperons and antihyperons. The relevant results of the CERN experiment NA57 are shown in Fig. 12.4. One can see that *the strangeness enhancement factor* for multi-strange hyperons, and antihyperons, is larger than for $|S| = 1$ particles, reaching 6–10 for $\Xi(\bar{\Xi})$, and about 20 for $\Omega(\bar{\Omega})$. A very similar pattern has been observed at RHIC.

A great success of the statistical-thermal model which correctly predicts relative particle yields[b] in relativistic nuclear collisions suggests, however, that instead of speaking of "strangeness enhancement" in nuclear collisions relative to elementary reactions, one should rather speak of "strangeness suppression" in the latter ones. In a central collision of relativistic nuclei, a thermally equilibrated state is formed. If for some reaction a relative

[b]The model has only two parameters: temperature T, and baryonic chemical potential μ_B. See Chapter 8 for more details.

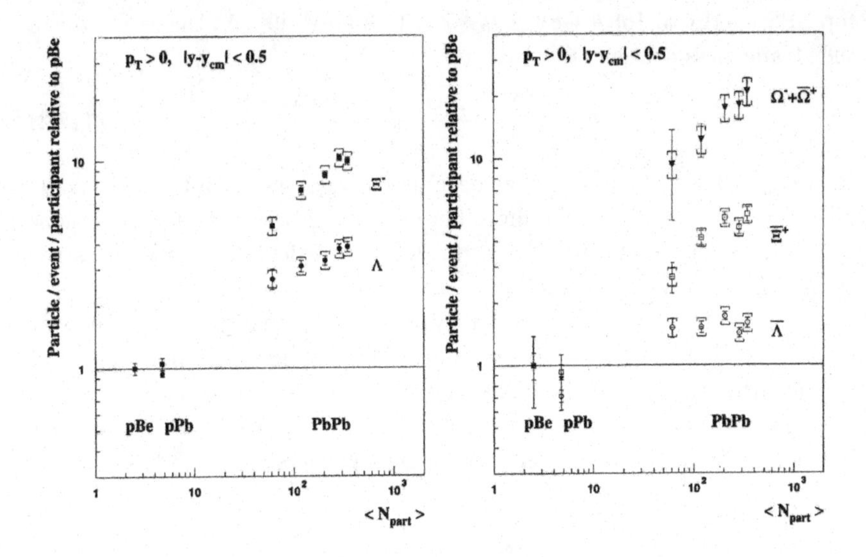

Fig. 12.4 Hyperon and antihyperon production in p+Be, p+Pb, and Pb+Pb collisions at 158A GeV (from Ref. [8]).

strangeness suppression is observed, this means that a full statistical equilibration has not been reached.

In Fig. 12.5 the relative strangeness content is shown as a function of the number of nucleons participating in the collision ("wounded nucleons",

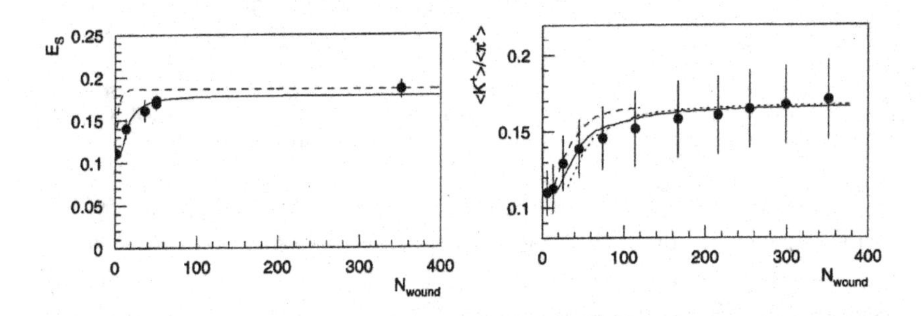

Fig. 12.5 System size dependence of strangeness production at SPS and RHIC energies ([7], data from Refs. [9] and [10]).

N_{wound}) at SPS energies (left panel), and at RHIC energies (right panel). Changing the number of N_{wound} in the left panel is achieved by studying central collisions of various nuclei (p+p, C+C, Si+Si, S+S, Pb+Pb) at the top SPS energy of 158A GeV ($\sqrt{s_{NN}} = 17.3$ GeV) [9]. As the estimator of the strangeness content, the quantity

$$E_s = \frac{\langle \Lambda \rangle + 2(\langle K^+ \rangle + \langle K^- \rangle)}{\langle \pi \rangle} \tag{12.3}$$

with $\langle \pi \rangle = 3/2(\langle \pi^+ \rangle + \langle \pi^- \rangle)$ has been used. Data shown in the right panel have been obtained by selecting different centralities in Au+Au collisions at $\sqrt{s_{NN}} = 200$ GeV [10]. Here the K^+/π^+ ratio has been used as the estimator of strangeness content.

In spite of these differences (let us also note some difference in the vertical scale), both plots are very similar. The strangeness content increases steeply with the size of the collision system up to about 60 participating nucleons, where it begins to saturate. Figure 12.6 shows that the same pattern is obtained for different strangeness carriers.

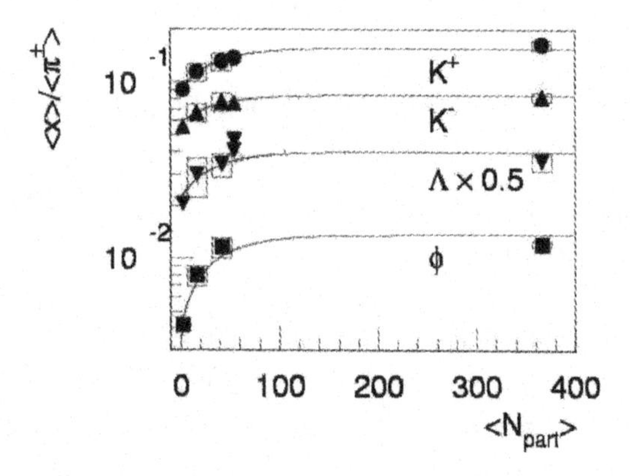

Fig. 12.6 System size dependence of strangeness production in Pb+Pb collisions at 158A GeV shown separately for four different strangeness carriers (from Ref. [11]).

Taking the viewpoint presented above, the relevant parameter for describing the evolution of the production of strangeness in relativistic heavy ion collisions would be *the strangeness undersaturation factor*, γ_s, defined as $\gamma_s \leq 1$, with $\gamma_s = 1$ for the full statistical equilibrium. To quote an example, $\gamma_s = 0.51$ for p+p collisions at $\sqrt{s} = 27.4$ GeV [12].

12.2 Heavy flavours

Investigation of the production of heavy flavours (charm and beyond) is much more difficult because of the very short lifetime of particles containing heavy quarks ($10^{-12}-10^{-13}$) s. This, and also typically multiparticle decay channels, make them not easily accessible for detection. Main properties of some particles carrying charm and beauty are listed in Table 12.2. At

Table 12.2 Charmed and beauty particles.

Particle	Mass, MeV	$c\tau$, μm	Dominant decay channels[a]
$D^{\pm}$	1869	312	$\bar{K}^0X, K^0X, K^{\pm}X$
D^0, $\bar{D}^0$	1864	123	$\bar{K}^0X, K^0X, K^{\pm}X$
Λ_c	2286	60	$NX, \Lambda X$
$B^{\pm}$	5279	491	$\bar{D}^0X, D^0X$
B^0, $\bar{B}^0$	5279	459	$K^{\pm}X, \bar{D}^0X, D^-X$
Λ_b	5624	369	

[a]Here "X"stands for "anything", which in most cases is a multiparticle final state.

SPS energies the relevant production cross sections are very small, and only some indirect evidence for an enhanced charm production in nuclear collisions has been obtained from the analysis of mass spectra of dileptons below the mass of the ρ-meson. For "elementary" reactions the dilepton mass spectrum in this interval can be understood as resulting from decays of a "cocktail" of various particles (including charmed ones) being produced with the known cross sections. In collisions of nuclei, however, a relative enhancement in this interval of the mass spectrum is observed, which can be partly explained if an enhanced production of charmed particles is assumed.

At RHIC the production cross sections are substantially larger, but at present no detector is equipped with a "vertex detector" with spatial resolution high enough as to record decay vertices of such short-lived particles. Direct reconstruction of $D^0(\bar{D}^0)$ mesons has been attempted by the STAR Collaboration in d+Au and Au+Au collisions at $\sqrt{s_{NN}} = 200$ GeV, using the decay channels $D^0 \to K^-\pi^+$ and $\bar{D}^0 \to K^+\pi^-$, the branching ratio being 3.8%. Figure 12.7 shows the invariant mass distribution for minimum bias Au+Au collisions. A clear D meson signal can be seen above the huge combinatorial background, which is due to very high multiplicity of analyzed events. The small statistics does not allow for any detailed analysis

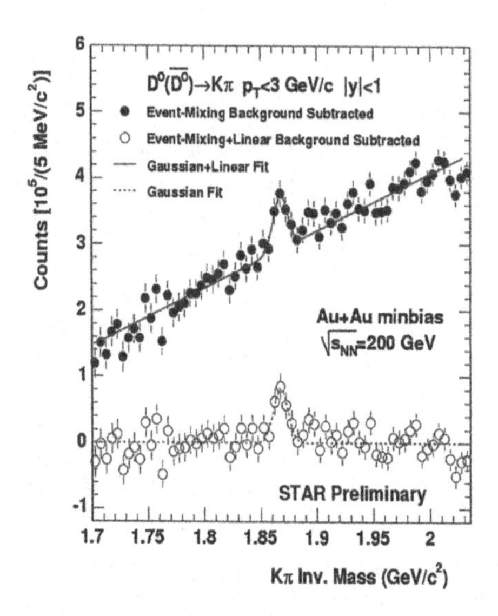

Fig. 12.7 $K\pi$ invariant mass distribution in the $D^0(\bar{D}^0)$ region from minimum bias Au+Au collisions at $\sqrt{s_{NN}} = 200$ GeV (from Ref. [13]).

of charm production, the data being restricted to low values of p_T.

More information can be obtained from the study of single electrons and muons coming from the also relatively rare semileptonic decays, analyzed by the PHENIX Collaboration. Electrons with high transverse momentum, $p_T \geq 1$ GeV/c, are believed to be of non-photonic origin (i.e. not resulting from $\gamma \to e^+e^-$ process, or Dalitz decays $\pi^0 \to e^+e^-, \eta \to e^+e^-$), and can be attributed to leptonic decays of particles containing heavy quarks. Electrons with $1 \leq p_T \leq 2$ GeV/c should come from D mesons, those with higher transverse momenta, $p_T \geq 4$ GeV/c, could be attributed to particles containing the b-quark.

Results from STAR show that the cross section for the production of charm at RHIC energies reaches the value of about (1.3 ± 0.2) mb per nucleon-nucleon collision, similar to that in p+p collisions. Figure 12.8 shows these results for various colliding systems at $\sqrt{s_{NN}} = 200$ GeV. Charm production seems to follow the binary scaling all the way from p+p to Au+Au collisions, what suggests that charm is predominantly produced at an early stage of the collision, and the contribution of secondary production processes is small. The perturbative QCD calculation underestimates the experimental results. More accurate data can be expected when the upgrade of RHIC detectors is completed. New data should also resolve

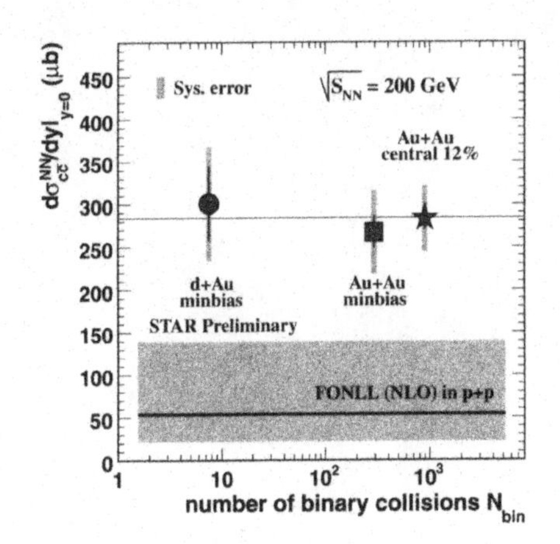

Fig. 12.8 Charm cross section at midrapidity per binary collision for d+Au, minimum bias Au+Au, and central Au+Au collisions at $\sqrt{s_{NN}} = 200$ GeV. Results of perturbative QCD calculation are shown with the grey band (from Ref. [14]).

the discrepancy between measurements of STAR and PHENIX (the charm cross sections from PHENIX are lower by about a factor of ~ 2 [13]).

References

[1] J. Rafelski and B. Müller, *Phys. Rev. Lett.* **48** (1982) 1066 P. Koch, B. Müller and J. Rafelski, *Phys. Rep.* **142** (1986) 167.
[2] M. Kowalski (NA35 Collaboration), *Nucl. Phys. A* **544** (1992) 609c.
[3] W. M. Yao *et al.* Particle Data Group, *J. Phys. G: Nucl. Part. Phys.* **33** (2006) 1.
[4] J. Podolanski and R. Armenteros, *Phil. Mag.* **45** (1954) 13.
[5] NA49 Collaboration, private communication.
[6] A. K. Wróblewski, *Acta Phys. Polon. B* **16** (1985) 379.
[7] C. Höhne, F. Pühlhofer and R. Stock, *Phys. Lett. B* **640** (2006) 96.
[8] A. Dainese *et al.* (NA57 Collaboration), *Nucl. Phys. A* **774** (2006) 51.
[9] C. Alt *et al.* (NA49 Collaboration), *Phys. Rev. Lett.* **94** (2005) 052301.
[10] S. S. Adler *et al.* (PHENIX Collaboration), *Phys. Rev. C* **69** (2004) 034909.
[11] C. Höhne (NA49 Collaboration), *Nucl. Phys. A* **774** (2006) 35.
[12] F. Becattini, *Z. Phys. C* **69** (1996) 485; F. Becattini and U. Heinz, *Z. Phys. C* **76** (1997) 269.
[13] A. A. P. Suaide, *J. Phys. G: Nucl. Part. Phys.* **34** (2007) S369.
[14] C. Zhong (STAR Collaboration), *J. Phys. G: Nucl. Part. Phys.* **34** (2007) S741.

Chapter 13

Emission of Light Nuclei, Antinuclei, and Hypernuclei

13.1 Light nuclei and antinuclei

In nuclear collisions, light nuclei emitted at rapidities close to that of beam or target are fragments of the colliding nuclei (see Chapter 7). Light nuclei, and also antinuclei, are, however, being emitted also in the central kinematic region, far from the beam or target rapidity. The dominant mechanism for this is believed to be *final state coalescence.*[a] Nucleons, or antinucleons, which have found themselves close in phase space (in close proximity and with small relative momenta) may form a nucleus or an antinucleus. A simple theory [1], based on probability arguments, gives the following relation between the yield of nuclei of mass A and momentum p and the yield of nucleons of momentum p/A

$$E\frac{d^3\sigma_A}{d^3p} = B_A\left(E\frac{d^3\sigma_N}{d^3(p/A)}\right)^A \tag{13.1}$$

where B_A is the *coalescence parameter* which characterizes the likelihood of the formation of a bound state of A nucleons.

Relation Eq. (13.1) has been verified at Bevalac and AGS energies. In Fig. 13.1 results from AGS are shown. Yields of light nuclei up to $A = 7$ in Au+Pb collisions at $11.5A$ GeV have been measured at rapidity $y = 1.9$. Over almost ten orders of magnitude the yields are well described by a simple exponential dependence. This dependence is very steep: adding a nucleon to a cluster generates a "penalty factor" of about 48.

It has also been found that invariant yields of various nuclei are proportional to their spin weight factor, $2J + 1$, and increase slightly with the binding energy per nucleon.

[a]Coalescence means "uniting into a whole".

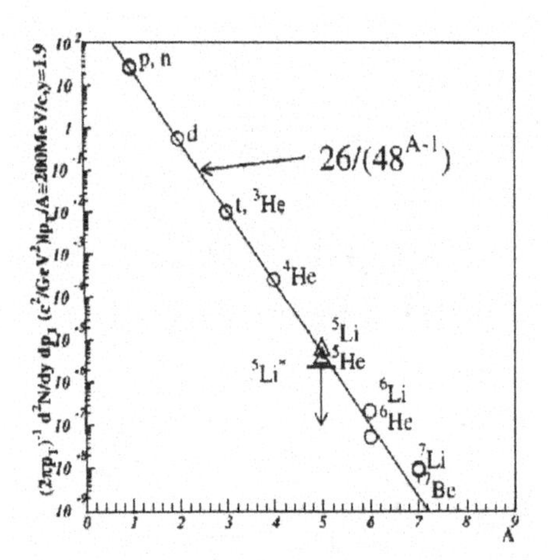

Fig. 13.1 Invariant yields of various nuclei from Au+Pb collisions at 11.5A GeV near $y = 1.9$ as a function of their mass number A (from Ref. [2]).

As the coalescence parameter, B_A, decreases with increasing energy of the collision (see below), at SPS and RHIC energies only the lightest nuclei could be detected. On the other hand, at SPS energies coalescence studies could have been extended also to light antinuclei.[b] Figure 13.2 shows the results of the CERN NA52 experiment which studied Pb+Pb collisions at 158A GeV. In the central region light nuclei and antinuclei with mass numbers $A \leq 3$ (d, $\bar{\text{d}}$, t, $\bar{\text{t}}$, ^{3}He, and $^3\overline{\text{He}}$) have been detected,[c] while nuclei with $A \leq 7$, and no antinuclei, have been detected in the fragmentation regions. A similar study at RHIC showed $\bar{\text{d}}$ and $^3\overline{\text{He}}$ [5].

Extraction of the coalescence parameters B_A and $B_{\bar{A}}$ from the data is not straightforward as the reference yields of protons and antiprotons should be corrected for the feed-down from hyperon and antihyperon decays. Due to uncertainties in this procedure the extracted numbers have relatively large errors. A compilation of coalescence parameters B_2 and B_3 is given in Fig. 13.3. The values extracted from the yields of nuclei and of antinuclei are compatible within experimental errors. An overall decrease of coalescence parameters with increasing energy of the collision is clearly seen.

In a statistical model which assumes thermal and chemical equilibrium

[b] Antideuterons were detected at the AGS, but the statistics was very low [3].

[c] Let us note similar yields of tritium and ^{3}He, as it might be expected for the coalescence mechanism.

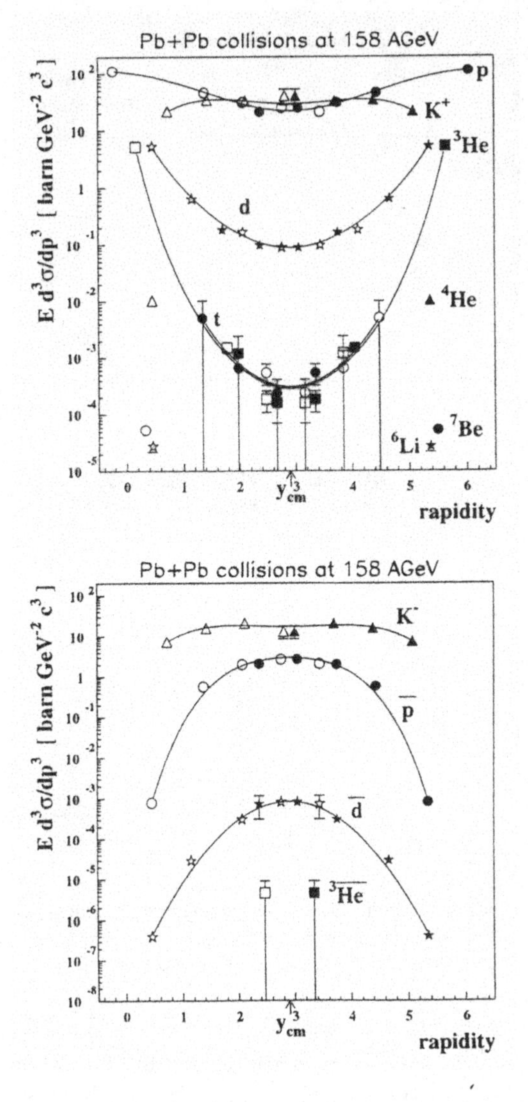

Fig. 13.2 Yields of secondary particles, antiparticles, light nuclei and antinuclei from Pb+Pb collisions at 158A GeV (from Ref. [4]).

in the "fireball" of a certain volume V, the coalescence parameter is related to this volume: $B_A \propto V^{-(A-1)}$ [1, 6]. For deuterons this means simply $B_2 \propto V^{-1}$. Extracting the size of the volume from the coalescence parameters is somewhat model-dependent, but the obtained values of the radius

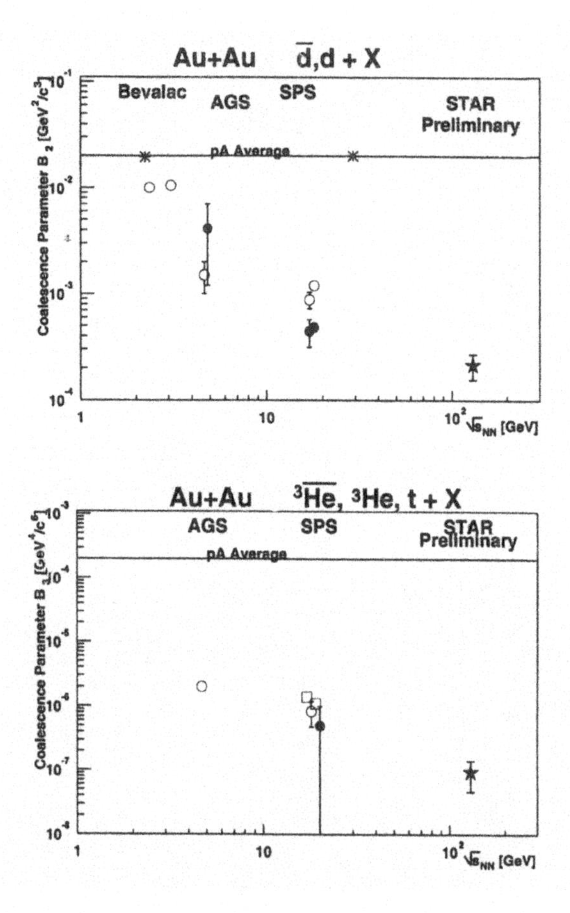

Fig. 13.3 Coalescence parameters B_2 (upper panel) and B_3 (lower panel) as functions of the collision energy for nuclei (hollow symbols) and antinuclei (full symbols). Also plotted are the average values for p+A collisions. From Ref. [5].

of the particle source are compatible with those obtained from particle interferometry (see Chapter 14). Coalescence studies may thus be considered as a complementary method to determine the size of the particle emission source. As it is known that the size of the fireball increases with increasing energy of the collision, a decrease of coalescence parameters with increasing energy, seen in Fig. 13.3, can be qualitatively understood.

If the particle emission volume is thermally and chemically equilibrated, then the invariant yields should obey the Boltzmann statistics

$$E\frac{d^3\sigma}{d^3p} \propto E\exp\left[-(E-\mu)/T\right] \tag{13.2}$$

where μ is the relevant chemical potential. For baryons $\mu = \mu_B$, and for antibaryons $\mu = -\mu_B$. Thus from the nucleus/antinucleus yield ratios at central rapidity the ratio μ_B/T can be derived [7]

$$E\frac{d^3\sigma_A}{d^3p_A}/E\frac{d^3\sigma_{\bar{A}}}{d^3p_{\bar{A}}} = \exp\left(2A\,\mu_B/T\right) \tag{13.3}$$

Taking the value of the "temperature" T from the slope of the spectra (see Chapter 10), the baryonic chemical potential μ_B can be evaluated. The value obtained from the $\mathrm{d}/\bar{\mathrm{d}}$ ratio ($A = 2$ in Eq. (13.3)) agrees quite well with that from the $\mathrm{p}/\bar{\mathrm{p}}$ ratio ($A = 1$ in Eq. (13.3)). This confirms the validity of the assumption that light nuclei and antinuclei are being formed by the coalescence mechanism in a thermally and chemically equilibrated fireball.

13.2 Hypernuclei

Hypernuclei, earlier called "hyperfragments", are nuclei which, apart of nucleons, contain hyperons (Λ or Σ). Such structures are not stable. Hypernuclei were discovered in 1952 by Danysz and Pniewski [8] in nuclear emulsions irradiated by cosmic rays in a stratospheric balloon flight. The physics of hypernuclei is mainly related to low-energy hyperon-nucleon interaction and to nuclear structure, but also to basic weak interactions. Extensive research on hypernuclei has been done using low-energy K^- beams from accelerators. Many hypernuclei have been identified and their binding energy measured. Figure 13.4 shows the binding energy, B_Λ, of the Λ hyperon in a hypernucleus as a function of its mass. This dependence is almost linear, and for $A \geq 10$ the values of B_Λ exceed the average binding energy of nucleons in similar nuclei which is about 8 MeV.

Hypernuclei can decay via the "mesonic" or "non-mesonic" modes — these are schematically shown in Fig. 13.5. In the "mesonic" decay mode a π^- meson is being emitted, while the "non-mesonic" mode results from the process $\Lambda + N \rightarrow N + N$, N standing for a proton or a neutron. Light hypernuclei decay via the "mesonic mode". The momentum of a proton from the "mesonic" $\Lambda \rightarrow \mathrm{p} + \pi^-$ decay is, however, only 100 MeV/c, which is below the nuclear Fermi level, and in heavy nuclei this process is blocked by the Pauli exclusion principle, as all nuclear energy levels are already occupied. On the contrary, nucleons from the "non-mesonic" mode have momenta of about 400 MeV/c, well above the Fermi level, and thus it is this decay mode which dominates in heavy hypernuclei. It should be pointed

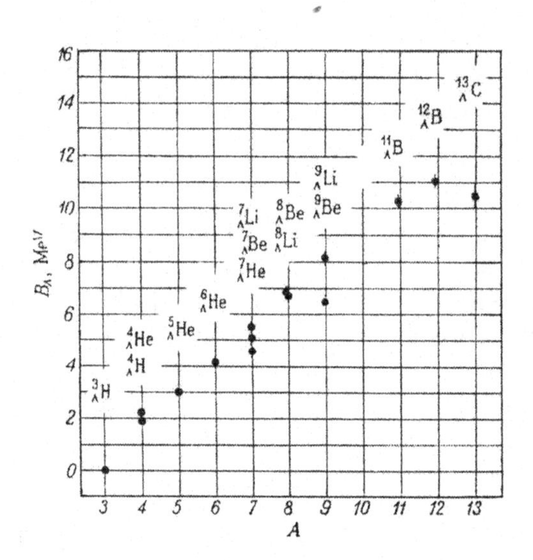

Fig. 13.4 Binding energy of the Λ hyperon in a hypernucleus as a function of its mass.

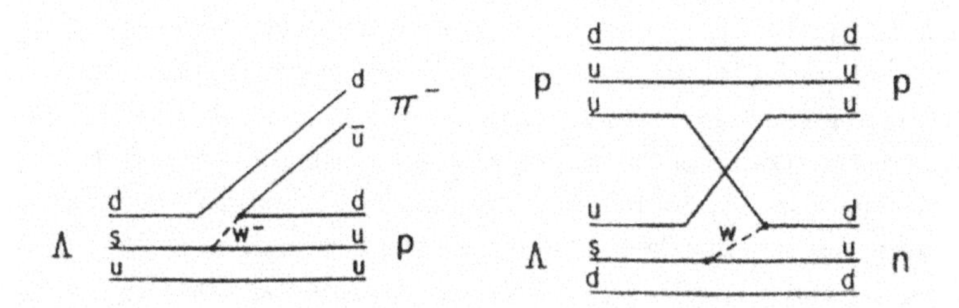

Fig. 13.5 Quark diagrams responsible for the "mesonic" (left) and "non-mesonic" (right) decay modes of hypernuclei.

out that "non-mesonic" decays of hypernuclei provide the unique possibility to study the four-fermion interaction.

"Double" hypernuclei with two bound Λ hyperons have also been reported [9, 10]. They are the only source of information about the Λ–Λ interaction, which appears to be attractive.

Measurements of the lifetime of hypernuclei are very interesting from the theoretical viewpoint. Due to the interaction with surrounding nucleons, the lifetime of a Λ hyperon in nuclear matter should be reduced. According to Dalitz [11] the lifetime of heavy hypernuclei should be 2–3 times shorter than that of a free Λ hyperon. Such measurements are difficult in "classical" experiments in which hypernuclei are produced in the target by a K^- beam, as hypernuclei emerge from the reaction as slow fragments and their path length in a detector is very short.

In 1974 Podgoretsky [12] and Okonov [13] suggested to produce relativistic hypernuclei using beams of relativistic nuclei. Hypernuclei can be formed from the incident nuclei by peripheral interactions, such as shown schematically in Fig. 13.6. One of the nucleons of the projectile nucleus

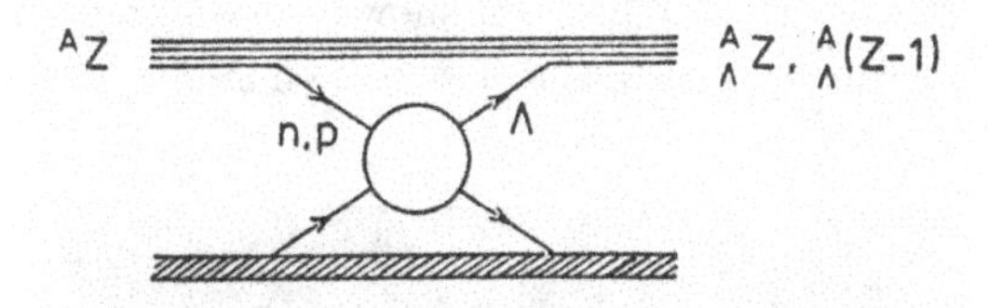

Fig. 13.6 Diagram of the peripheral production of a hypernucleus in a nucleus-nucleus collision.

${}^AZ$ interacts with a nucleon in the target, producing a Λ hyperon which is then captured in the projectile forming a hypernucleus with mass number A and atomic number Z or $Z-1$ (for light projectiles both should be about equally probable). High momentum, and the Lorentz boost, lengthen the path of hypernuclei produced in such a process, providing suitable conditions for a measurement of their lifetime. The first experiment of this kind was performed in 1975/76 at LBL [14]. The beam of 2.1A GeV ^{16}O ions was focused on a polyethylene target. The detector consisted of wide-gap spark chambers and scintillation counters. The event trigger required that a heavy ion interacted in the target and a K^+ decayed 11 ns or later after the primary interaction. Spark chambers were photographed and searched for tracks originating from a vertex outside the target. Twenty-two events were attributed to hypernuclear decays occuring between the target and the spark chambers. They supposedly represent a mixture of ${}^{16}_{\Lambda}$O and ${}^{16}_{\Lambda}$N hypernuclei, produced mainly in three-body reactions ${}^{16}\mathrm{O} + \mathrm{p} \to {}^{16}_{\Lambda}\mathrm{O} + \mathrm{n} + K^+$ and ${}^{16}\mathrm{O} + \mathrm{n} \to {}^{16}_{\Lambda}\mathrm{N} + \mathrm{n} + K^+$. The obtained lifetime estimate for a mass 16 hypernucleus was $(0.86^{+0.33}_{-0.26}) \times 10^{-10}$ s, shorter than that of a free Λ, and the production cross section per target nucleon $\sigma = (2 \pm 1)\mu$b.

Ten years later a similar experiment was performed in Dubna, using the 3.7A GeV ^{4}He and 3.0A GeV ^{7}Li beams, and also a polyethylene target [15]. A streamer chamber placed in the magnetic field of 0.9 T, and a set of scintillation counters served as detectors. The event trigger was set to detect an ion with one more unit of charge behind the chamber as compared to that in front of it, what would match a hypernucleus decay in the chamber with the emission of a π^- meson. The latter should also leave a track in the streamer chamber. A number of events have been attributed to decays of ${}^4_\Lambda$H, and also some ${}^7_\Lambda$Li have been identified. The measured lifetime of the ${}^4_\Lambda$H hypernucleus was $(2.6 \pm 0.6) \times 10^{-10}$ s, compatible with that of a free Λ, but the production cross section per target nucleon was found to be $\sigma \cong 0.03\,\mu$b. much lower than that obtained in [14], but compatible with theoretical predictions [16]. Figure 13.7 shows a photograph of the decay of a relativistic ${}^4_\Lambda$H hypernucleus in the streamer chamber. Figure 13.8 shows

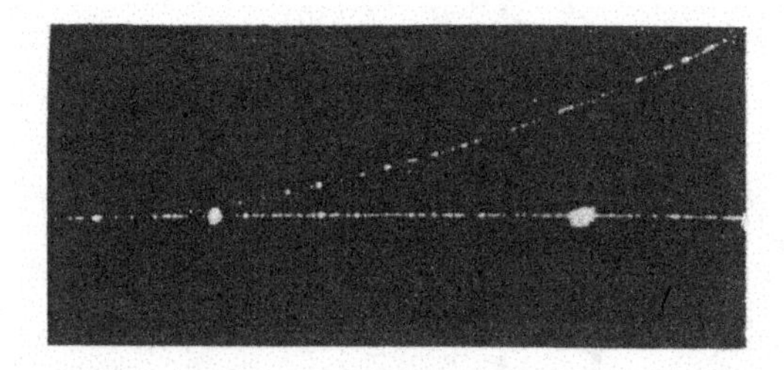

Fig. 13.7 Decay of a relativistic hypernucleus ${}^4_\Lambda\text{H} \rightarrow {}^4\text{He} + \pi^-$ recorded in a streamer chamber in Dubna (from Ref. [15]).

a compilation of the measured lifetimes of hypernuclei. The two full points have been obtained in experiments in which relativistic hypernuclei were produced [14, 15]. In addition, one should quote an interesting result from a devoted experiment at GSI, in which the lifetime of the Λ hyperon in very heavy ($A > 180$) hypernuclei was obtained as $\tau = (1.45 \pm 0.11) \times 10^{-10}$ s [17]. This would confirm the theoretical prediction [11].

Light hypernuclei should, however, also be produced by the coalescence mechanism, together with light nuclei and antinuclei. They should be searched for in the central kinematic region. Hypernuclei produced "thermally" in collisions of relativistic nuclei have been reported by the E864 experiment at the AGS [2]. Among secondary particles produced in Au+Pb collisions at 11.5A GeV the ${}^3_\Lambda$H hypernuclei have been identified from a signal in the invariant mass $M({}^3\text{He}, \pi^-)$ (these particles result from the dominant "mesonic" decay channel ${}^3_\Lambda\text{H} \rightarrow {}^3\text{He} + \pi^-$).[d] Within the frame

[d]In the E864 detector decay vertices occuring at a distance of a few cm from the target could not be seen.

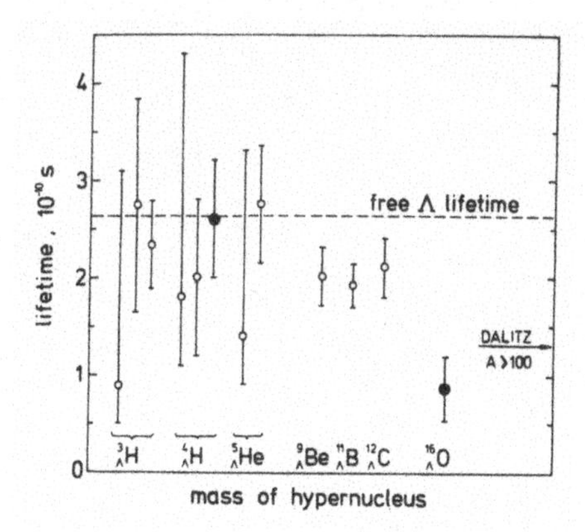

Fig. 13.8 Compilation of hypernuclear lifetimes. The two full points are from experiments with relativistic heavy ion beams [14, 15].

of the coalescence model, the probability of the formation of $^3_\Lambda$H should be comparable to that of ^{3}He (values of the relevant coalescence parameters B_3 should be similar), but it turns out that the probability of the first process is several times lower. Thus, coalescing a Λ hyperon instead of a nucleon seems to generate an additional "penalty factor", what may be related to the "undersaturation of strangeness" at AGS energies (see Chapter 12).

References

[1] A. Z. Mekjian, *Phys. Rev. C* **17** (1978) 1051.
[2] L. E. Finch *et al.* (E864 Collaboration), *Nucl. Phys. A* **661** (1999) 395c.
[3] G. Van Buren (E864 Collaboration), *Nucl. Phys. A* **661** (1999) 391c.
[4] S. Kabana *et al.* (NA52 Collaboration), *Nucl. Phys. A* **638** (1998) 411c.
[5] D. Hardtke *et al.* (STAR Collaboraction), *Nucl. Phys. A* **698** (2002) 671c.
[6] L. P. Csernai and J. I. Kapusta, *Phys. Rep.* **131** (1986) 223.
[7] G. Ambrosini *et al.* (NA52 Collaboration), *Phys. Lett. B* **417** (1998) 202.
[8] M. Danysz and J. Pniewski, *Phil. Mag.* **44** (1953) 348.
[9] M. Danysz *et al.*, *Phys. Rev. Lett.* **11** (1963) 29; *Nucl. Phys.* **49** (1963) 121.
[10] D. J. Prowse, *Phys. Rev. Lett.* **17** (1966) 782.
[11] R. H. Dalitz, *Proc. Int. Conf. Hyperfragments*, ed. W. O. Lock, St.-Cergue, CERN 64-1 (1963), p. 147.
[12] M. I. Podgoretsky, Report JINR–8309, Dubna (1974), p. 81.
[13] E. O. Okonov, Report JINR–8309, Dubna (1974), p. 104.
[14] K. J. Nield *et al.*, *Phys. Rev. C* **13** (1976) 1263.
[15] S. A. Avramenko *et al.*, *Nucl. Phys. A* **547** (1992) 95c.
[16] M. Wakai, *Nucl. Phys. A* **547** (1992) 89c.
[17] P. Kulessa *et al.*, *J. Phys. G: Nucl. Part. Phys.* **28** (2002) 1715.

Chapter 14

Hadronic Femtoscopy

Correlations of particles with close momenta are sensitive to the space-time characteristics of the emitting source. Thus studying such correlations one can obtain information about sizes of the order of femtometers ($1 \text{ fm} = 10^{-15}$ m). The relevant analysis procedures are called *hadronic femtoscopy*. We will discuss correlations of identical bosons and those of identical fermions separately, as the underlying physics differs in these two cases: boson correlations are mainly due to properties of their wave functions (*boson interferometry*), while for identical fermions correlations are mainly due to final state interactions. The latter is also true for the case of non-identical particles.

14.1 Correlations of identical bosons

Discovery, underlying physics and general formalism

The first experimental observation relevant to hadron interferometry was that of Goldhaber *et al.* [1] who found that in antiproton annihilations in hydrogen the distribution of the opening angle of pion pairs of the same charge ("like" pions) differs from that for opposite charge ("unlike" pions). The reaction studied was $\bar{\mathrm{p}} + \mathrm{p} \to n\pi^+ + n\pi^- + m\pi^0$ with $n = 2$ or 3 and $m = 0, 1, 2, \ldots$ at 1.05 GeV/c incident antiproton momentum. The "like-charge" pion pairs tend to be emitted with relatively smaller opening angle, *i.e.* closer to each other in phase space. The observed effect was just of opposite sign to that of Coulomb interaction which causes the repulsion of like-charge particles. In the subsequent paper [2] an explanation of this effect was given, based on symmetrization of the wave function for identical bosons. The relevant mathematical derivation will be given below, but the effect could be qualitatively understood if one remembers that pions obey Bose-Einstein statistics which enhances the probability of finding more than one particle in any given quantum state.[a]

Soon after, similar correlations were observed between charged pions produced in pion–proton collisions, and it was found that the effect comes mainly from pairs of pions with close momenta [3]. Observation of correlations between neutral pions [4] confirmed the interpretation as a result of Bose–Einstein statistics. Some years later, this early evidence was completed by similar findings for strange bosons: $K_s^0 K_s^0$ pairs [5], and K^+K^+ and K^-K^- pairs [6].

Also, it has been realized that the correlation pattern is sensitive to the size of the emitting source. Several years after observation and understanding the Bose-Einstein correlations between identical pions, it has been noticed that similar considerations were already developed earlier in astronomy for photons[b] coming from stellar objects by Hanbury-Brown–Twiss [7]. They invented the "intensity interferometer" which measures the average product of intensities (not amplitudes!), and showed that the intensity correlations can be used for the determination of sizes of stars. This apparent analogy, however, should not be taken strictly, as in fact there are basic differences between the two situations. In particle and/or nuclear physics the distance between the particle emission points (sources) is much smaller that the distance between the detectors, while in astronomy the opposite situation occurs [8]. The discussed method of boson interferometry is often being referred to as Hanbury-Brown–Twiss, or HBT, method, but, according to Ref. [9] this should rather be avoided.

A simple theoretical derivation goes along the following lines. Let us consider two identical bosons with momenta $\vec{k}_1$ and $\vec{k}_2$ emitted from space points $\vec{r}_A$ and $\vec{r}_B$ of an extended source with density distribution $\rho(\vec{r})$ with $\int d^3\vec{r}\,\rho(\vec{r}) = 1$. If their wave functions are described by plane waves, then the amplitude for detecting the first boson at point $\vec{r}_1$ and the second one at $\vec{r}_2$ is in general given by

$$A(\vec{k}_1, \vec{k}_2) = \frac{1}{\sqrt{2}}[e^{i\vec{k}_1(\vec{r}_1 - \vec{r}_A)}e^{i\vec{k}_2(\vec{r}_2 - \vec{r}_B)} + e^{i\vec{k}_1(\vec{r}_1 - \vec{r}_B)}e^{i\vec{k}_2(\vec{r}_2 - \vec{r}_A)}] \tag{14.1}$$

[a] At very low temperatures this property of Bose-Einstein statistics ultimately leads to *Bose–Einstein condensate* in which all bosons occupy the same (lowest) energy level, with their wave functions fully overlapping.

[b] Photons have spin 1 and thus are also bosons.

The presence of the two terms follows from the basic principles of quantum mechanics which requires the amplitude for a given process to be taken as the sum of amplitudes corresponding to all possible ways of reaching the final state. In other terms, one should perform summation over all trajectories between initial and final state points. In our example there are

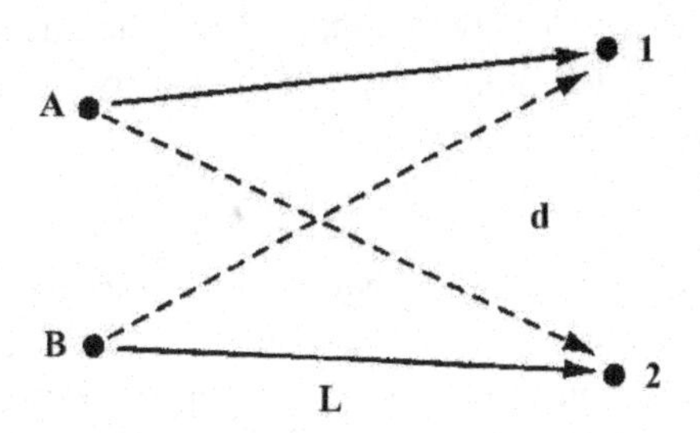

Fig. 14.1 Alternative paths of particles emitted from points A and B of the source, and reaching detectors situated at points 1 and 2.

two indistinguishable ways of reaching the final state which are shown in Fig. 14.1 with continuous and dashed lines, respectively, and Eq. (14.1) just reflects this situation. Assuming that boson production amplitudes have random phases (a "chaotic" source) one can perform the integration and obtain the expression for the probability of detecting two such bosons

$$W(\vec{k}_1, \vec{k}_2) = \int d^3\vec{r}_A \, d^3\vec{r}_B |A(\vec{k}_1, \vec{k}_2)|^2 \rho(\vec{r}_A) \, \rho(\vec{r}_B) \tag{14.2}$$

This "double" probability can be identified with the two-particle correlation function which is defined as the ratio of the two-particle density to the product of two single-particle densities

$$C_2(\vec{k}_1, \vec{k}_2) = \frac{\langle n \rangle^2}{\langle n(n-1) \rangle} \frac{d^6\sigma/d\vec{k}_1 d\vec{k}_2}{d^3\sigma/d\vec{k}_1 \, d^3\sigma/d\vec{k}_2} \tag{14.3}$$

For distances L (between the source and the detectors) and d (between the detectors) both much larger than the source radius R, this correlation function depends only on the vector momentum difference, or relative momentum, $\vec{q} = \Delta\vec{k} = \vec{k}_1 - \vec{k}_2$, and thus can simply be written as

$$C_2(\vec{q}) = 1 + f(\vec{q}) \tag{14.4}$$

where $f(\vec{q})$ is the square of the Fourier transform of the source density distribution $\rho(\vec{r})$, usually normalized to one. The relative momentum $\vec{q}$ is conjugate under Fourier transform to the space distance $\vec{r}_A - \vec{r}_B$ between the emission points. Various relative momentum components will then be conjugate under Fourier transform to various spatial distances, and thus one can expect that by choosing suitable variables information can be obtained not only about the mean size, but also about the geometrical shape of the emitting source.

This approach can be easily generalized to four dimensions by replacing three-momenta in the above given formulae by four-momenta. The correlation function is then written as

$$C_2(\vec{q}, q_0) = 1 + f(\vec{q}, q_0) \tag{14.5}$$

with $q_0 = E_1 - E_2$ being the energy component which, under Fourier transform, is conjugate to the time duration of the emission process. Thus the correlation function in four-momenta is the Fourier transform of the space-time structure of the source. Neglecting the time dependence corresponds to the specific assumption of a static source, without correlation between production points and momenta.

The correlation function C_2 exhibits a peak at small values of $|\vec{q}|$, the width of this peak being inversely proportional to the size of the source, as shown in Fig. 14.2. From Eqs. (14.4) and (14.5) it follows that at $|\vec{q}| = 0$

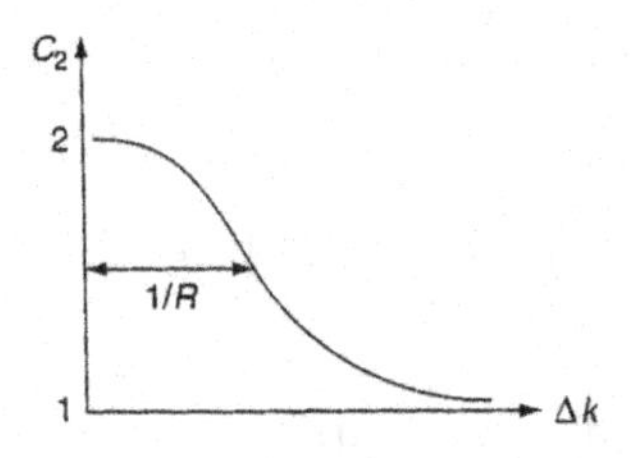

Fig. 14.2 Shape of one-dimensional correlation function for two identical bosons.

the correlation function might reach the value of 2, but in reality it never does so. This would mean that the emitting source may be only partly chaotic and partly coherent, the coherent emission not contributing to the correlation peak near $|\vec{q}| = 0$. In order to account for this, in Ref. [10] a phenomenological parameter λ was introduced into the correlation function

$$C_2(\vec{q}, q_0) = 1 + \lambda f(\vec{q}, q_0), \quad 0 \leq \lambda \leq 1 \tag{14.6}$$

The parameter λ is often called the chaoticity (or incoherence) parameter, its correct name, however, should rather be the *correlation intensity parameter*. This is because also other effects, both physical such as production of intermediate-state resonances, and methodical such as admixture of particles of different type to the studied sample would also reduce the value of λ.[c] Thus the experimentally determined value of λ gives only a lower limit for the degree of chaoticity of the source. In fact, the chaoticity can be larger, the apparent correlation peak being reduced by other effects.

In order to obtain a more detailed description of the source, a certain form of the density distribution should be assumed, and suitable variables chosen for the parameterization of the correlation function.

Static source

In their pioneering works, Kopylov and Podgoretsky [11], and also Cocconi [12], considered a uniformly radiating disc emitting with a characteristic time τ. For the correlation function they obtained the expression

$$C_2(q_T, q_0) = 1 + \lambda \frac{2J_1(q_T R)/q_T R}{1 + (q_0 \tau)^2} \tag{14.7}$$

where q_T is the projection of the vector $\vec{q}$ onto the transverse momentum plane (see Fig. 14.3), and J_1 is the first-order Bessel function.

Yano and Koonin [13] and Gyulassy, Kaufmann and Wilson [14] pro-

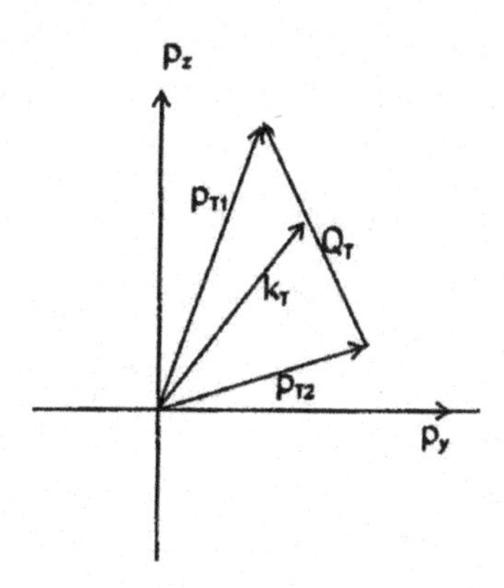

Fig. 14.3 Diagram showing the relative momentum components in the transverse momentum plane (the plane perpendicular to the collision axis).

[c] If the studied sample of like-charge particles contains a fraction f_1 of bosons of a certain type (*e.g.* pions), and a fraction f_2 of bosons of another type (*e.g.* kaons), and possibly also some fraction $1 - f_1 - f_2$ of non-bosons (*e.g.* muons or electrons), then the value of the correlation intensity parameter λ would not exceed $f_1^2 + f_2^2$.

posed a Gaussian parameterization of the source space-time density

$$\rho(\vec{r},t) = C\,\exp(-r^2/R^2 - t^2/\tau^2) \tag{14.8}$$

which gives the correlation function

$$C_2(q,q_0) = 1 + \lambda\,\exp(-q^2R^2/2 - q_0^2\tau^2/2) \tag{14.9}$$

This formula can be easily generalized to an ellipsoidal source characterized by different radii in different directions. It seems plausible to assume a cylindrical geometry, i.e. azimuthal symmetry around the collision axis, and consider an ellipsoid elongated (or contracted) along this axis. For such density function

$$\rho(\vec{r},t) = C\,\exp[-(x^2+y^2)/R_T^2 - z^2/R_L^2 - t^2/\tau^2] \tag{14.10}$$

the correlation function is[d]

$$C_2(q_T,q_L,q_0) = 1 + \lambda\,\exp(-q_T^2R_T^2/2 - q_L^2R_L^2/2 - q_0^2\tau^2/2) \tag{14.11}$$

where q_T and q_L are, respectively, the transverse and longitudinal components of the vector momentum difference between the two bosons, and R_T and R_L are the transverse and longitudinal radii.

To complete the review of various theoretical approaches one should mention that by Weiner [15], based on quantum optics. In the picture of partial coherence, assuming superposition of coherent and chaotic fields, the correlation function should be the sum of two coupled Gaussians with effective radii differing by $\sqrt{2}$:

$$C_2(q_T) = 1 + 2p(1-p)\,\exp(-q_T^2R_T^2/2) + p^2\,\exp(-q_T^2R_T^2) \tag{14.12}$$

This gives a somewhat more peaked structure than a single Gaussian, a feature which is actually observed in some experiments. The parameter p, called "chaoticity", represents the fraction of particles emerging from the chaotic source: $p = \langle n_{\rm cha}\rangle/(\langle n_{\rm cha}\rangle + \langle n_{\rm coh}\rangle)$. There are, however, some basic objections to this approach [16]. The argument is that the number of particles in a quantum coherent state is not defined (can be any number, with an appropriate weight factor), and thus for charged particles coherent states cannot exist because of charge conservation. There seems to be no objections against using this picture for neutral bosons such as π^0's.

[d] The factor 1/2 appearing in all terms of the exponent in this formula, similarly as in Eq. (14.9), is sometimes being omitted for simplicity. This, however, rescales the obtained values of the radii by the factor $1/\sqrt{2}$.

Expanding source

That particle interferometry can give information about the space-time evolution of the system was already noticed in some early works. Pratt [17] considered a spherical, radially expanding source, while Kolehmainen and Gyulassy [18] introduced the correlations between the dynamics and geometry assuming that formation time of a particle increases linearly with its energy. They have written the correlation function in the form

$$C_2(q_T, \Delta y, m_{T1}, m_{T2}) = 1 + \lambda |G(p_1, p_2)|^2 / [G(p_1, p_1)\, G(p_2, p_2)] \quad (14.13)$$

where the function G is

$$G(p_1, p_2) = a\, K_0(\sqrt{u})\, \exp(-q_T^2 R_T^2 / 4) \quad (14.14)$$

where $K_0(\sqrt{u})$ is the modified Bessel function of the complex argument which includes the transverse masses m_{T1}, m_{T2} ($m_T = \sqrt{m^2 + k_T^2}$), rapidity difference Δy, proper time τ_0, and the temperature T of the source. The value of this temperature should be assumed, or obtained from the data, e.g. from the single-particle transverse momentum spectra, what makes this analysis rather complicated.

More recently, another set of variables has been introduced by Bertsch [19], also for a system with cylindrical geometry. These are $q_{T\text{out}}$ (often

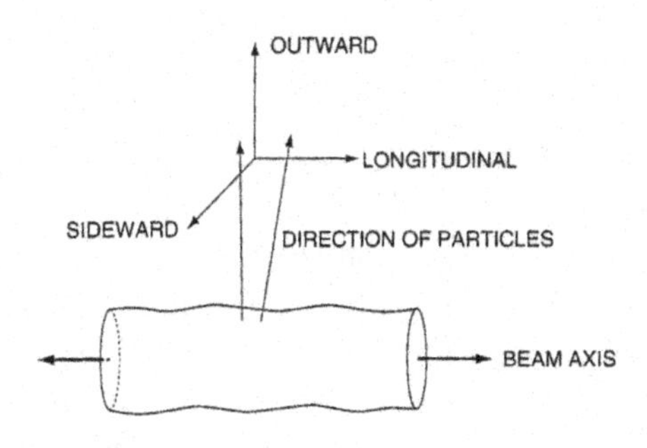

Fig. 14.4 The Bertsch coordinates.

simply called q_{out} or q_O), $q_{T\text{side}}$ (called q_{side} or q_S), and q_{long} (called q_L). With $\vec{q} = \vec{k}_1 - \vec{k}_2$ and $\vec{p} = \vec{k}_1 + \vec{k}_2$, they are defined as follows (see Fig. 14.4):

$$q_S \perp p_T, \perp p_L$$

$$q_O \parallel p_T, \perp p_L$$

$$q_L \parallel p_L, \perp p_T$$

Bertsch argues that by using these variables one can obtain more information about the collision process, namely q_S is related to the transverse dimension of the source, R_T, q_L is related to the time of longitudinal expansion τ_f, and the difference $(q_O - q_L)$ to the time duration of particle emission $\Delta\tau_f$. Thus one can get insight into the space-time evolution of the system, and e.g. distinguish between the two cases: whether the system emits particles within a very short time after a quick evolution, or the system evolves slowly and the emission is also spread over a longer time. It is the latter case which might be expected if a quark-gluon plasma is formed in a nuclear collision.

The Bertsch correlation function can be written as

$$C_2(\vec{q}) = 1 + \lambda \exp(-R_S^2\, q_S^2 - R_O^2\, q_O^2 - R_L^2\, q_L^2) \tag{14.15}$$

It was later improved by adding an interference term between q_O and q_L

$$C_2(\vec{q}) = 1 + \lambda \exp(-R_S^2\, q_S^2 - R_O^2\, q_O^2 - R_L^2\, q_L^2 - 2\, R_{OL}^2 q_O q_L) \tag{14.16}$$

For a longitudinally boost-invariant source the cross-term R_{OL} vanishes.

Expansion of the source leads to the m_T-dependence of radius parameters: the radii R_L and R_S decrease with an increasing velocity of a pion pair [20]. This dependence for R_L is of the form

$$R_L \propto \sqrt{\frac{T}{m_T}} \tag{14.17}$$

while that for R_S is slightly weaker

$$R_S \propto \frac{1}{\sqrt{(1 + m_T \beta_T / T)}} \tag{14.18}$$

where β_T is the transverse expansion velocity.

Before reviewing the experimental data, it might be worthwhile to discuss invariant properties of the correlation function. As the square of four-momentum is Lorentz-invariant, the correlation function $C_2(q^2)$ is also Lorentz-invariant, what is sometimes accentuated by explicitly calling this variable q_{inv}. Using the decomposition of $\vec{q}$ into transverse and longitudinal components, one obtains the corresponding radii R_T (invariant) and R_L (non-invariant). For the emission volume being a tube along the collision axis, and for an observer placed at midrapidity y_0 of the collision, one obtains $R_L(y) \sim \cosh(y - y_0)$ where y is the rapidity of the boson pair, while for an observer sliding along the y-axis the shape of the emission volume will be correctly reproduced, yielding $R_L(y) = \text{const}$.

Background evaluation

In application of intensity interferometry to experimental data one uses correlation function normalized to the "reference" or "background" distribution, with the symbol C_2 retained for this new function

$$C_2(\vec{q}, q_0) = A_2(\vec{q}, q_0)/B_2(\vec{q}, q_0) \tag{14.19}$$

where $A_2(\vec{q}, q_0)$ is the distribution of pion pairs when both pions come from the same event, $B_2(\vec{q}, q_0)$ is the "random" distribution of pion pairs formed using pions from different events, selected in the same way (or just belonging to the same sample). The B_2 distribution should include all experimental correlations contained in A_2 except for those due to Bose–Einstein correlations. Thus the correlation function C_2 measures the influence of Bose-Einstein correlations on the distribution of momentum differences of like-charge pion pairs. Some other methods of background generation as *e.g.* using like-charge pions from the same event, but with momentum components reshuffled in order to destroy correlations, have also been tried and shown to yield almost the same results. Unlike-charge pion pairs can also be used for the background distribution, but there might be differences in π^+ and π^- production rates, and the resonances in $(\pi^+\pi^-)$ system could also distort the distributions.

Coulomb correction

Dealing with charged particles one should not forget the Coulomb forces acting between them, and also between the investigated pair and the emitting source. The standard correction is the *Gamov factor*, R_G, derived for point-like charges in the non-relativistic approximation. It depends on their relative momentum q

$$R_G(\vec{p}_1, \vec{p}_2) = R_G(q) = \frac{2\pi\eta}{\exp(2\pi\eta) - 1} \tag{14.20}$$

with $\eta = \alpha m/q$, and $\alpha = 1/137$. This correction is important only for small q, as R_G approaches unity for large q. It is also more important for heavier particles. For pions $2\pi\eta \cong 7$ MeV $/q$, and the correction is often being neglected, for kaons $2\pi\eta \cong 25$ MeV $/q$. The simple Gamov correction cannot be really adequate for nuclear collisions. Thus in Ref. [21] it has been proposed to use pions of opposite signs of charge ($\pi^+\pi^-$ pairs) to evaluate the role of Coulomb interaction (for unlike-charge pairs there is no contribution of the Bose–Einstein symmetrization to the correlation function). Figure 14.5 shows the correlation functions for $\pi^+\pi^-$ pairs together with the Gamov curve, and with phenomenological functions calculated for finite-size sources. One can see that the Gamov function is far off the data, and the final size of the pion emitting source should be taken into account. Assuming that the functions representing Coulomb attraction and Coulomb

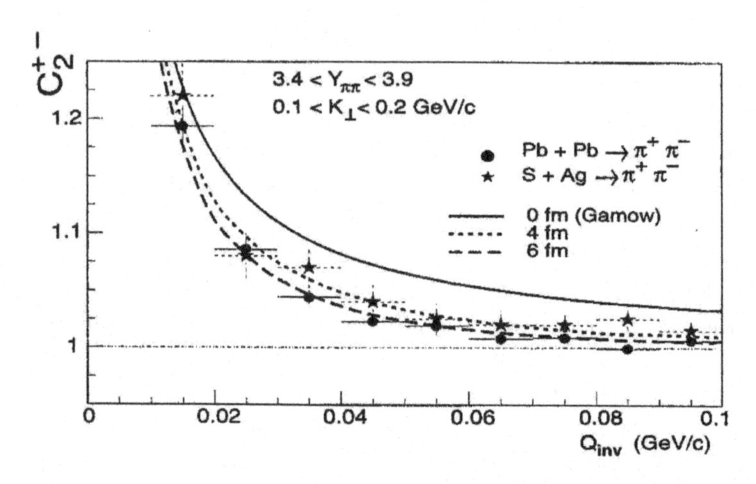

Fig. 14.5 Correlation functions for $\pi^+\pi^-$ pairs from central collisions of S+Ag at 200A GeV and Pb+Pb at 158A GeV. Curves show the standard Gamov correction factor and phenomenological Coulomb functions calculated for finite-size sources with radii 4 fm and 6 fm (from Ref. [21]).

repulsion do not differ, one can use the inverse of the function found for unlike-charge pairs as Coulomb correction for like-charge pairs.

$$C^{--}_{\text{Coulomb}} = C^{++}_{\text{Coulomb}} = (C^{+-})^{-1} \tag{14.21}$$

It has been found that this phenomenological correction describes much better the experimental data.

The central potential (interaction of the pair with the residual source) is important only for nuclear collisions, where a residual source being a many-nucleon system might have a high charge. This interaction does not introduce any correlations, but distorts the q-scale, leading to the inequality

$$R^{\text{apparent}}_{++} \leq R^{\text{source}} \leq R^{\text{apparent}}_{--} \tag{14.22}$$

The differences are, however, only at a few percent level.

Results for elementary collisions

Boson interferometry has been extensively used for elementary reactions. Nearly all results obtained to date are for two-pion measurements, mainly for $\pi^-\pi^-$. Most of the published data come from track chamber experiments, mainly bubble and streamer chambers, in which event statistics was never very high. Data with higher statistics come from experiments at the CERN colliders: the ISR for pp collisions and LEP for e^+e^- collisions. In low-statistics experiments only one-dimensional analysis could be performed, and most groups used either the Kopylov–Podgoretsky parameterization with $q_0 = 0$ or a single Gaussian in q. Only a few groups studied correlations in two dimensions using either Kopylov-Podgoretsky or Gaussian. The result, in general, was $R \approx 1\,\text{fm}$ and $\tau \approx 1\,\text{fm/c}$, as expected for elementary collisions, both values determined with relatively large errors. The correlation intensity parameter λ strongly varies between various groups, but it is this parameter which is the most sensitive to various experimental biases and corrections introduced to the data. A few attempts to detect the non-spherical shape of the pion source indicated its slight elongation along the collision axis ($R_T/R_L \approx 0.6$–0.8).

K^+K^+ correlations have also been studied, yielding $R_{KK} < R_{\pi\pi}$, what was initially interpreted as evidence for kaon emission occuring at an earlier stage of the expansion of the source. However, it has been argued that a decrease of R with increasing particle mass can be derived from the uncertainty principle. In Ref. [22] a conjecture was made that as the maximum of Bose–Einstein correlations occurs when two bosons are very close in phase space (the relative momentum $\vec{q} \to 0$), the uncertainty principle should have some effect here. It provides the two well-known relations:

$$\Delta p \cdot \Delta r \simeq \hbar \tag{14.23}$$

and

$$\Delta E \cdot \Delta t \simeq \hbar \tag{14.24}$$

Simplifying slightly the derivation presented in Ref. [22] let us replace in the first relation Δp by the particle momentum p, and Δr by the distance between the two particles, r, and in the second relation ΔE by the particle energy, $E = p^2/2m$. Then from the first uncertainty relation one obtains

$$r \simeq \hbar/p \tag{14.25}$$

and from the second one

$$p \simeq \sqrt{2m\hbar/\Delta t} \tag{14.26}$$

Combining the last two expressions one obtains

$$r \simeq \sqrt{\hbar \cdot \Delta t/2m} \tag{14.27}$$

Assuming that Δt is the characteristic time scale of strong interactions, independent of the particle identity and of its mass, one finally obtains

$$r \simeq \text{const.}/\sqrt{m} \tag{14.28}$$

It turns out that this formula describes well the measured mass dependence of the source size.

Finally, it should be mentioned that genuine three-particle correlations have been observed for pions produced in high energy e^+e^- collisions in LEP. In analogy to pair correlation studies, the parameterization used was

$$C_3 = 1 + 2\lambda_3 \exp(-R_3^2 Q_3^2) \tag{14.29}$$

where Q_3 is the invariant mass of a pion triplet.

Results for nuclear collisions

Boson interferometry has been also applied to nuclear collisions at relativistic energies. Here, of course, larger sizes could be expected due to large sizes of the colliding objects, and new features such as some dependence on the size and impact parameter (or degree of overlap) of colliding nuclei might become apparent. First experimental data were obtained at LBL, JINR Dubna and CEN Saclay. Due to limited statistics these data were analyzed in terms of a single radius parameter and were not subdivided into various subsamples. Their compilation is shown in Fig. 14.6. These data show compatibility of the "radius" estimated from the correlation function with radii of the colliding nuclei (or with that of the smaller nucleus in case of asymmetric collisions).

Figure 14.7 shows exemplary one-dimensional correlation functions for pions and kaons from Au+Au central collisions at 10.8A GeV. One can see that the correlation function for K^+K^+ pairs is wider than those for $\pi^+\pi^+$ or $\pi^-\pi^-$ pairs, meaning that $R_{KK} < R_{\pi\pi}$, similarly as it was found in elementary collisions. Radii for $\pi^+\pi^+$ and $\pi^-\pi^-$ pairs do not differ within experimental accuracy. The value of λ is about 0.5 for all three samples.

Improved detectors and increasing multiplicities in nuclear collisions at higher energies allow to use a multi-parameter method of analysis, such as that of Yano–Koonin–Podgoretsky (YKP), or that of Bertsch. A combined analysis of YKP correlation parameters, and of transverse spectra of charged secondary particles, performed for Pb+Pb collisions at 158A GeV within the frame of the expanding source model allowed to obtain the source parameters: the freeze-out temperature $T = (120 \pm 12)$ MeV and transverse expansion velocity $\beta_T = 0.55 \pm 0.12$. The graphical representation of this result is given in Fig. 14.8.

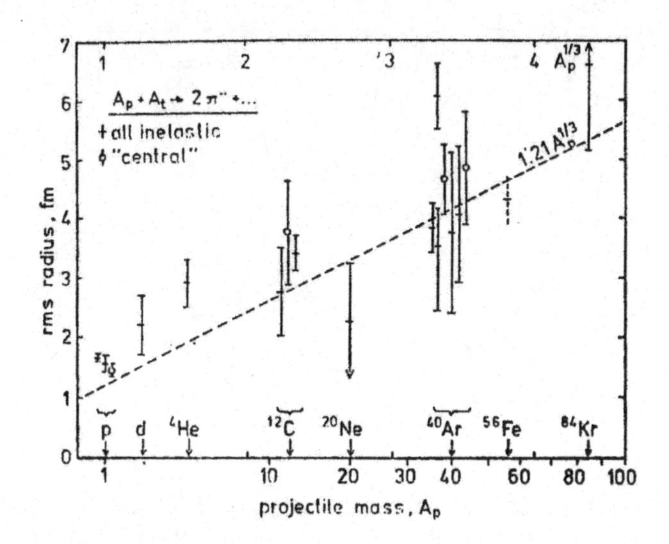

Fig. 14.6 Compilation of pion source radii obtained from one-dimensional correlation functions in the energy range up to $5A$ GeV (from Ref. [23]).

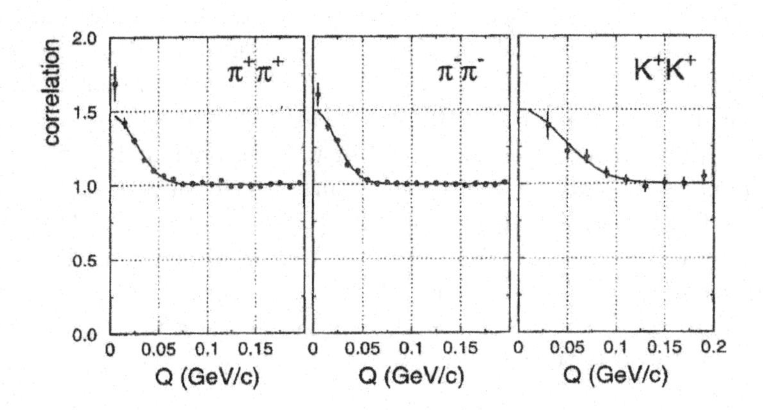

Fig. 14.7 One-dimensional two-meson Coulomb-corrected correlation functions for central Au+Au collisions at $10.8A$ GeV. Curves are Gaussian fits to the data (from Ref. [24]).

The Bertsch formalism has gradually become the main approach to the boson interferometry in collisions of relativistic nuclei. Figure 14.9 shows some early results from the CERN SPS showing an increase of all three R-parameters with increasing size of the colliding system. Among the R-

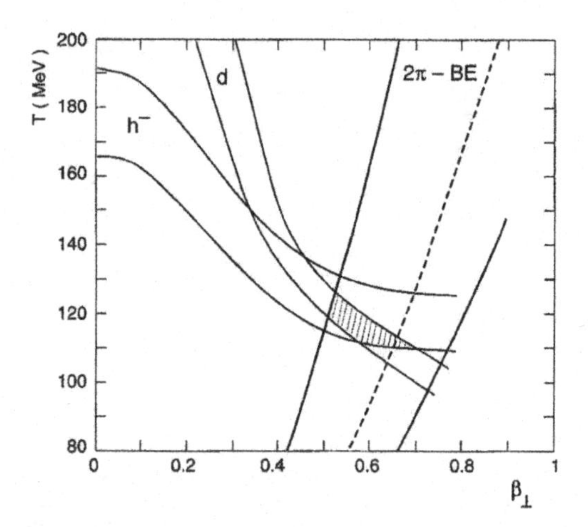

Fig. 14.8 Two-dimensional plot showing allowed regions of freeze-out temperature and of radial expansion velocity for central Pb+Pb collisions at 158A GeV near midrapidity, derived from fits to the two-pion correlation function and to transverse mass spectra. Bands are drawn at ± 1 st.dev. around fitted values. The common allowed region is shaded (from Ref. [21]).

parameters R_L is larger than R_O or R_S. Again, the kaon source radii are smaller, $R_{KK} < R_{\pi\pi}$, and the pion and kaon radii were found to show a common $1/\sqrt{m_T}$ scaling [26]. Larger radii for pions were qualitatively explained here by the contribution of resonances. This looks plausible as particles from resonance decays should show larger radii, and a large fraction of pions and only few kaons come from resonance decays.

Results of a detailed analysis of central Pb+Pb collisions at five energies covering the entire energy range of the SPS, are displayed in Fig. 14.10. These data show a decrease of the radii with increasing transverse momentum k_T of a pion pair, this decrease being the strongest for R_L. Also, for all energies we have the relation

$$R_L > R_O > R_S \tag{14.30}$$

A small positive value of the difference $R_O - R_S$ means a short emission time $\Delta\tau_f \cong$ (2–4) fm/c. The lifetime of the source, τ_f, is estimated to be about twice longer. A relatively large value of the longitudinal parameter R_L indicates a significant elongation of the emission region along the collision axis, and the fact that both transverse radii R_O and R_S are bigger than geometrical radii of the colliding nuclei suggests an important role of the transverse flow. Indeed, the *blast wave model*, fitted to single-particle p_T-

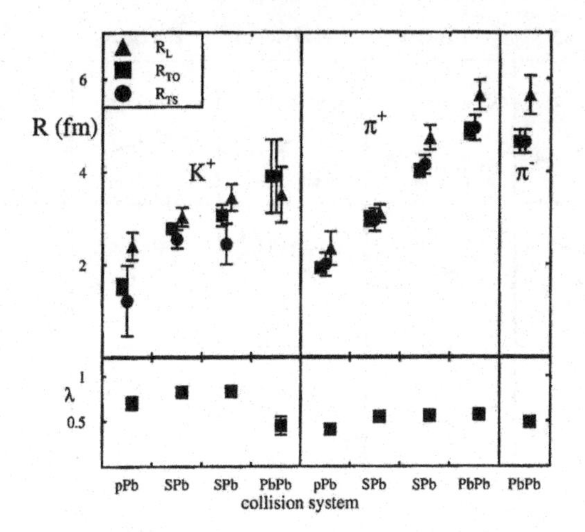

Fig. 14.9 The Bertsch correlation parameters R_L, R_O (denoted R_{TO}), and R_S (denoted R_{TS}) for kaon and pion pairs for different collision systems at SPS energies (from Ref. [25]).

spectra, gives at the same time a good description of the k_T-dependence of the radii — this is shown in Fig. 14.10 with continuous lines. The cross-term in Eq. (14.16) is compatible with zero at midrapidity.

A compilation of p+p, p̄+p, and heavy ion (AGS and early SPS) two-pion interferometry results has been attempted in Fig. 14.11. The source radii are seen to increase approximately as the cube-root of the multiplicity density dn/dy of charged particles at midrapidity. This behaviour indicates that the freeze-out occurs at a certain average (pion) density, what means that in an expanding system pions interact with their local environment only. This seems to be compatible with short-range interaction.

As R_S gives the transverse radius of the pion emitting source, and R_L is related to its longitudinal dimension, the product $R_S^2 R_L$ can be taken as an estimate of the volume of the source. In Fig. 14.12 this quantity is plotted as a function of the density of negatively charged particles at midrapidity, dn^-/dy, for several reactions studied at the SPS. The data are compatible with a linear rise of the volume with the number of particles produced in the collision.

At still higher energies, at the RHIC collider, the R-parameters do not change much as compared to their values at SPS energies, but they continue to grow slowly with dn_{ch}/dy, approximately proportionally to $(dn_{\text{ch}}/dy)^{1/3}$ [30]. This is shown in Fig. 14.13, together with data from lower energies.

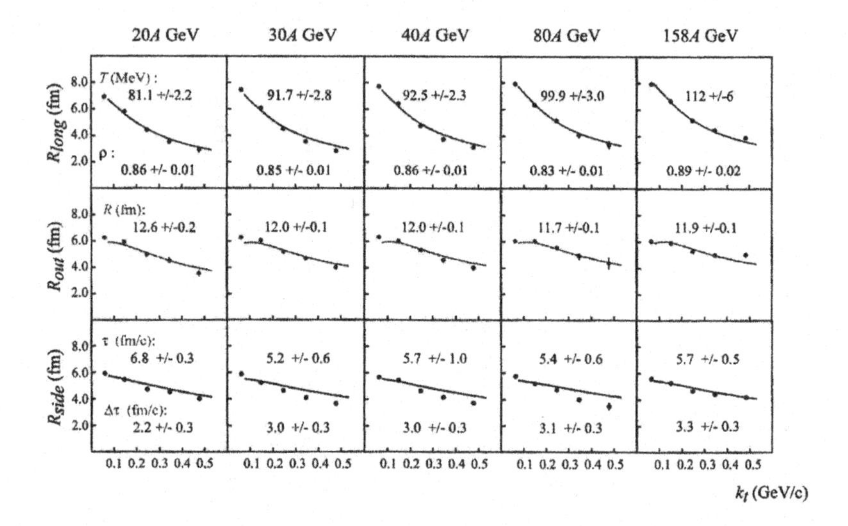

Fig. 14.10 The k_T-dependence of Bertsch parameters R_L, R_O, and R_S at midrapidity for central Pb+Pb collisions at the SPS, for five different energies. Lines correspond to a combined fit of the blast-wave model to the radii and to single-particle p_T spectra (from Ref. [27]).

The left panel shows the Bertsch radii for p+p, d+Au, Cu+Cu, and Au+Au collisions at RHIC as measured by the STAR Collaboration plotted against the cubic root of the multiplicity density of charged secondaries at midrapidity. The right panel shows radii as measured by PHENIX, compared to those measured at lower energies (AGS and SPS). An almost linear dependence of R_S and R_L on $(dn_{\rm ch}/dy)^{1/3}$ is visible, at least for the RHIC and SPS data. It implies that R_S and R_L, which are the "geometric" radii, follow the same dependence for different collisions over a wide range of energies. Thus the product $R_S^2 R_L$ should grow linearly with dn/dy, what again means that freeze-out occurs at a certain density of the final state. This is an important result which seems to be valid for final state systems containing predominantly pions, i.e. from the SPS energies (at lower energies there is a large fraction of baryons in the final state, even at midrapidity). The results from RHIC indicate, however, that the line drawn in Fig. 14.12 is slightly too steep.

Genuine three-pion correlations, already detected in some elementary processes, have been also observed in collisions of relativistic nuclei. Significant three-pion correlations have been seen in Pb+Pb collisions at SPS energies [31, 32], and also in Au+Au collisions at RHIC [33]. Here not the

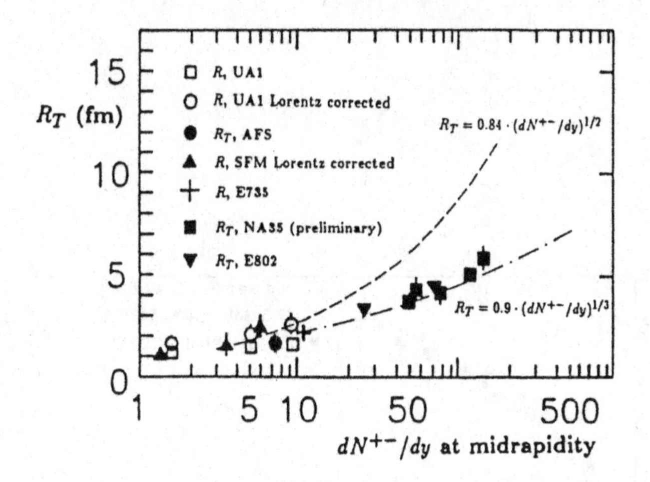

Fig. 14.11 Source radii from two-pion correlations as a function of charged particle midrapidity density (from Ref. [28]).

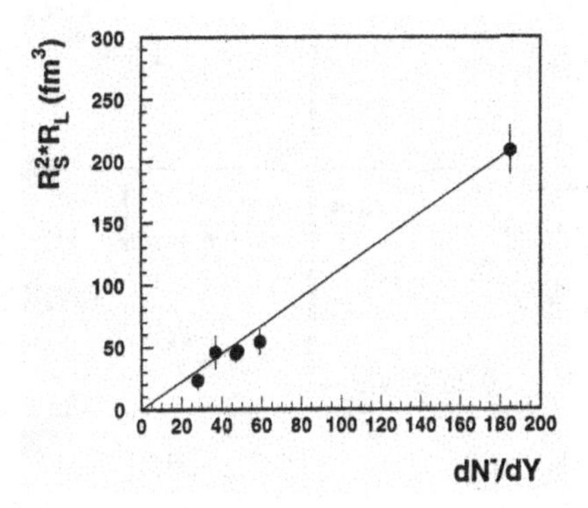

Fig. 14.12 The product $R_S^2 R_L$ plotted versus density of negatively charged particles at midrapidity, dn^-/dy, for nuclear collisions at the SPS. The four lower points are for S+S, S+Cu, S+Ag and S+Au collisions, the highest point is for Pb+Pb collisions (from Ref. [29]).

width of the correlation function, but the value of the intercept at $Q_3 = 0$ has been taken as a measure of the effect. This intercept, $r_3(Q_3 = 0)$ reaches the value of about one.

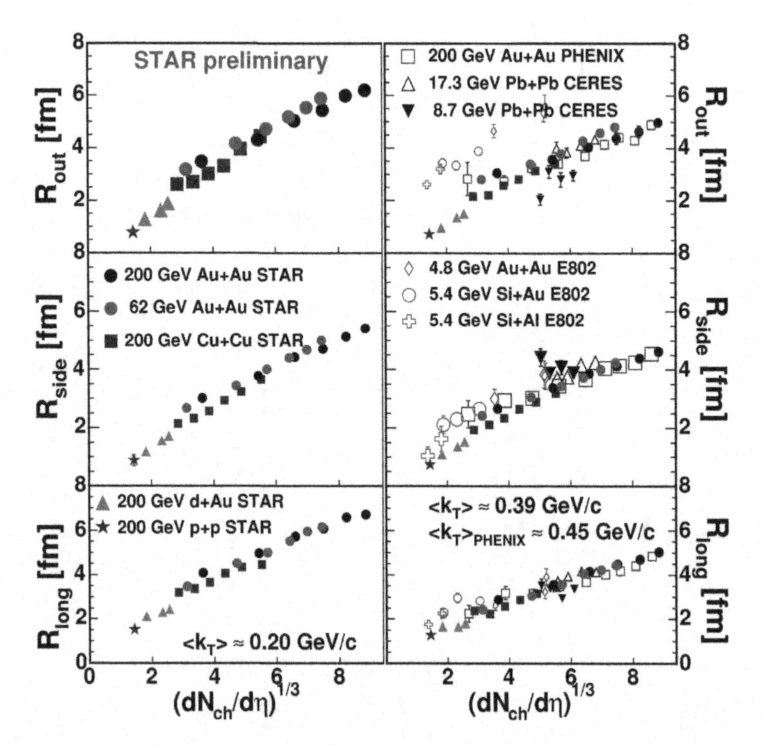

Fig. 14.13 The Bertsch radii for various collision systems and energies plotted against the cubic root of the multiplicity density of secondary charged particles at midrapidity (this is the corrected version of the plot from Ref. [30]).

14.2 Correlations of identical fermions

The correlation function for a pair of identical fermions differs very much from that for identical bosons. The wave function for identical fermions should be antisymmetric, and this leads to the suppression of the probability to find two fermions with close momenta (*the Pauli exclusion principle*). Thus it can be expected that the correlation function will have a minimum at the relative momentum $q = 0$. This simple conjecture remains true when the spin structure of a pair of identical spin $\frac{1}{2}$ fermions is taken into consideration: such a system can be in a singlet $s = 0$ or triplet $s = 1$ states with relative weights proportional to statistical spin factors $2J + 1$ what gives the ratio of 1:3 . It is only for a triplet state that the spatial part of the wave function is antisymmetric in order to combine with a symmetric spin part into an antisymmetric total wave function. As this state prevails by 3:1, the correlation function for $q \to 0$ will have a minimum. The region very close to $q = 0$ is, however, practically not accessible because of finite experimental resolution, and for larger values of q the shape of the correlation function is determined by final state interactions. Thus it is mainly these interactions, and not quantum interference phenomena, which can be studied with pairs of protons or lambda-hyperons.

Following Ref. [35][e] the correlation function can be written as

$$B(q,p) = A_c^{\pm}(k^*)[1 + B_0(q,p) + B_i(q,p)] - 1 \tag{14.31}$$

where $q = p_1 - p_2$, $p = p_1 + p_2$, and $k^* = \sqrt{-q^2}$. The contribution of the Coulomb interaction is determined by the value of the factor

$$A_c^{\pm}(k^*) = \pm\frac{2\pi}{k^* a_c}\left[\exp\left(\pm\frac{2\pi}{k^* a_c}\right) - 1\right]^{-1} \tag{14.32}$$

where a_c is the Bohr radius, the plus sign corresponds to repulsion, and the minus sign to attraction of the charges. $A_c(k^*)$ approaches unity for $k^* a_c/2\pi \gg 1$, but differs considerably from unity for $k^* a_c/2\pi \leq 1$. If the particles have single charges and equal masses, then $a_c = 2\hbar^2/me^2 = 274\hbar/mc$, and for protons $a_c = 57.5$ fm. From this one can see that the larger the mass m, the wider is the region of momenta k^* for which it is necessary to include the Coulomb interaction.

The term $B_0(q,p)$ describes the effect of quantum statistics. If the source distribution is Gaussian, then $B_0(q,p) = -\frac{1}{2}\exp\left(-r_0^2 q^2 - \tau_0^2 q_0^2\right)$,

[e] The treatment of Ref. [35] is more general than that of the earlier Ref. [34], in particular there is no assumption of simultaneous emission.

the expression similar to that for identical bosons, but with "minus" sign, and with R called r_0.

The term $B_i(q,p)$ describes the interaction in the final state (FSI), and can be calculated numerically from the known values of the scattering lengths for the particles studied.

For a pair of protons, the complete correlation function after a steep rise at small values of q, shows a broad maximum near $q = 20$ MeV/c. the height of which decreases with the increasing size of the source [34, 35]. For large sizes the correlation function becomes flat. The proton–proton correlation function, calculated for three different sizes of the source, characterized by the source radius, r_0, is shown in Fig. 14.14.

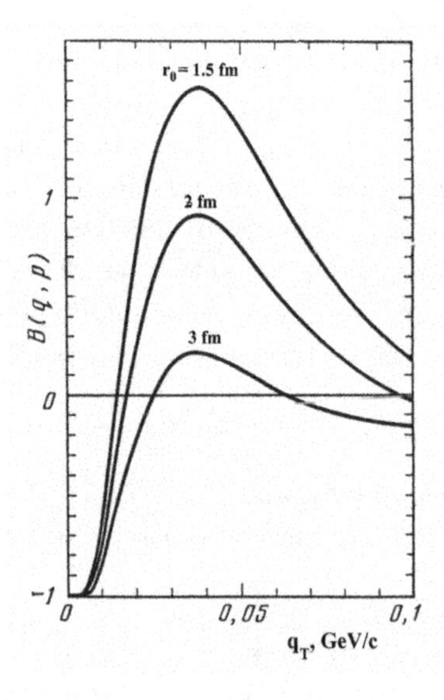

Fig. 14.14 Proton–proton correlation function calculated for three different values of the radius of the source (from Ref. [35]).

Correlations of protons emitted with small relative momenta were observed for the first time in $\pi^- + Xe$ interactions at 9 GeV/c [36]. In collisions of relativistic nuclei proton–proton correlations have been studied over a wide range of incident energies: from a few GeV (synchrophasotron [37, 38] and AGS [39, 40]), through SPS energies [41, 42], up to RHIC energies [43]. In Fig. 14.15 results for Au+Au central collisions from the AGS at four different energies are shown. The extracted source radii vary

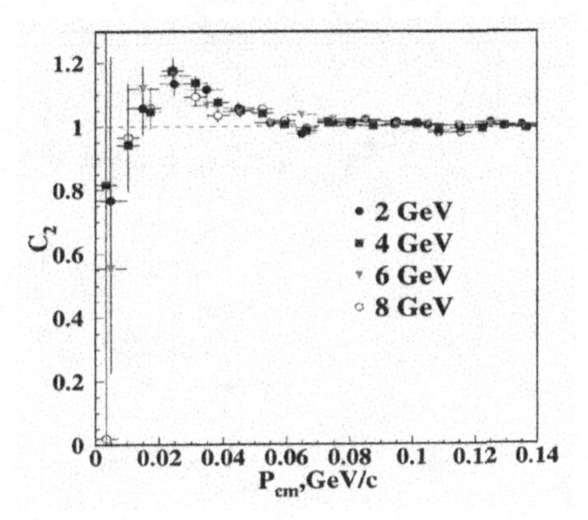

Fig. 14.15 Proton–proton correlation functions for Au+Au central collisions at four different beam energies (from Ref. [39]).

between 6 and 9 fm, depending on the used parameterization,[f] with no significant energy dependence. In Fig. 14.16 results for p+Pb and S+Pb collisions at 200A GeV are given. The values of the extracted radii are indicated in the Figure. Figure 14.17 shows analogous results for Pb+Pb central collisions at 158A GeV. Finally, in Fig. 14.18 results for Au+Au

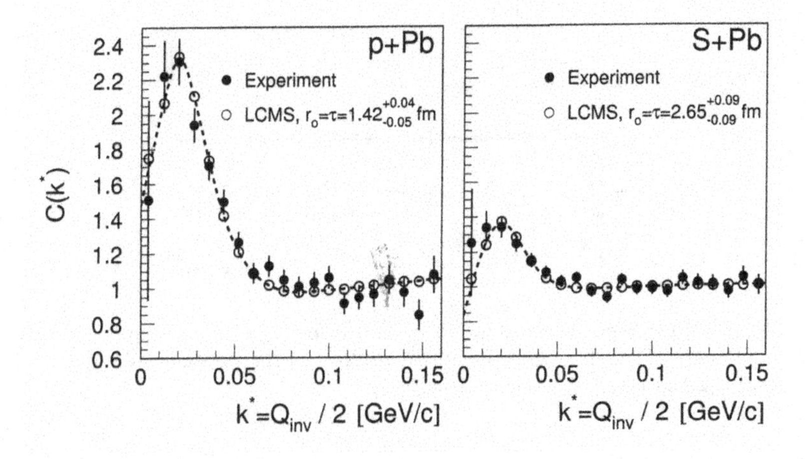

Fig. 14.16 Proton–proton correlation functions for p+Pb (left panel) and S+Pb (right panel) collisions at 200A GeV (from Ref. [41]).

[f] Let us recall that the radius of the gold nucleus is about 7 fm.

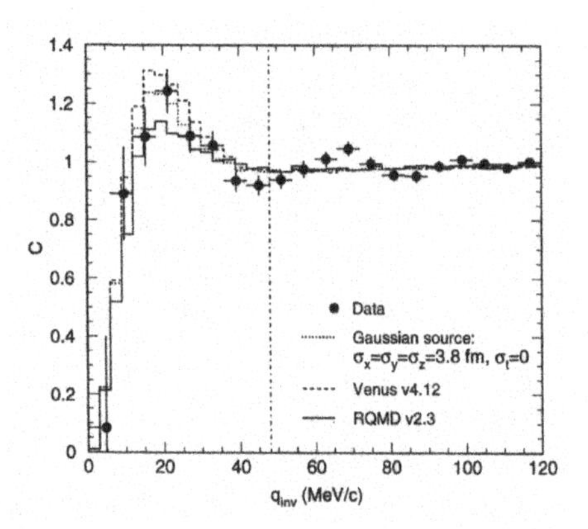

Fig. 14.17 Proton–proton correlation function for Pb+Pb central collisions at 158A GeV (from Ref. [42]).

collisions at $\sqrt{s_{NN}} = 200$ GeV with three different centrality selections are given. The fitted curves have been calculated according to Ref. [35].

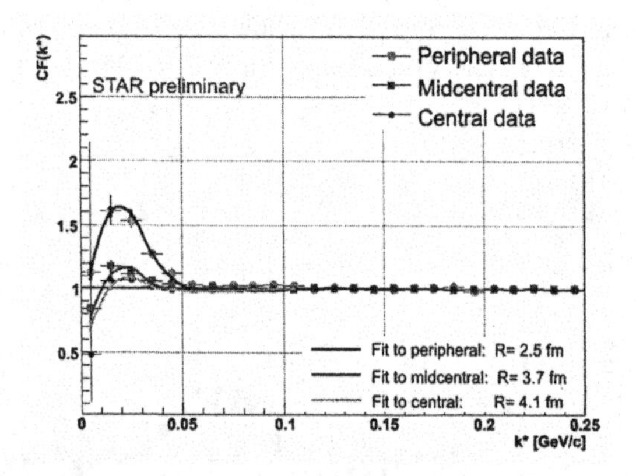

Fig. 14.18 Proton–proton correlation functions for Au+Au collisions at $\sqrt{s_{NN}} =$ 200 GeV with three diferent centralities. Values of the extracted Gaussian radii are indicated in the Figure (from Ref. [43]).

The general features shown by the data can be summarized as follows:

(i) the measured source sizes increase with the mass of the projectile, and in central collisions of heavy nuclei are comparable to their sizes;
(ii) no significant changes with increasing beam energy are observed;

(iii) there is some dependence on the centrality of the collision: measured size increases with increasing number of participating nucleons.

Results obtained from proton–proton correlation studies turn out to be consistent with trends established for other types of particles (pions, kaons): the source radii scale with the cubic root of event multiplicity, and follow the $1/\sqrt{m_T}$ dependence [43]. Antiproton–antiproton correlations, which have also been studied by the STAR Collaboration, look compatible with proton-proton correlations [43].

An attempt to measure Λ–Λ correlations in central Pb+Pb collisions at $158A$ GeV has also been reported [44]. The obtained distribution of Λ–Λ pairs shows the anticorrelation at small relative momenta due to the Fermi statistics, and indicates a relatively weak final state interaction.

14.3 Correlations of non-identical particles

While identical particle correlations reflect the properties of quantum statistics, and of the final state interactions (Coulomb and strong), non-identical particle pairs are sensitive to the final state interactions only. This gives access to some information on mutual interaction also between short-lived particles, for which no direct scattering experiments are feasible.

Correlations between $p(\bar{p})$ and $\Lambda(\bar{\Lambda})$ in various combinations in central Au+Au collisions at $\sqrt{s_{NN}} = 200$ GeV have been studied in Ref. [45]. Results for p–$\Lambda(\bar{p}$–$\bar{\Lambda})$ correlations confirmed the earlier measurements for p–Λ at AGS [46], and at SPS [47]. The $\bar{p}$–$\Lambda(p$–$\bar{\Lambda})$ correlations have been studied for the first time. Figure 14.19 shows the correlation functions for the two cases considered. Due to the absence of quantum statistics effects, and of the Coulomb interaction, their shapes differ very much from that for p–p correlations, and the two curves differ also between themselves. Radii extracted from p–$\Lambda(\bar{p}$–$\bar{\Lambda})$ and those from $\bar{p}$–$\Lambda(p$–$\bar{\Lambda})$ are different, the latter ones being significantly smaller. This might reflect, apart from the presence of annihilation channels, a difference in the corresponding scattering lengths, the result which would not be accessible by any other method.

Correlations between non-identical particles, similarly as for identical ones, are also sensitive to the space-time structure of the system at freeze-out and to its dynamical evolution. In particular, comparing the correlation functions with relative velocity parallel and anti-parallel to the pair velocity, one can infer the emission order and the time interval between the two

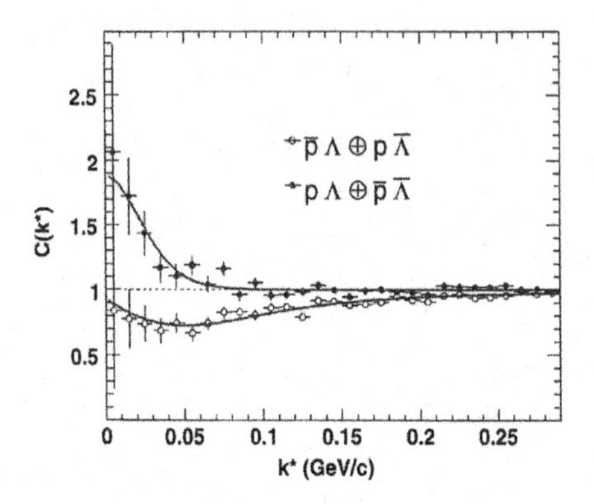

Fig. 14.19 Combined (p–Λ)⊕(p̄–Λ̄) and (p̄–Λ)⊕(p–Λ̄) correlation functions from central Au+Au collisions at $\sqrt{s_{NN}} = 200$ GeV. Fitted curves have been calculated according to Ref. [35]. The Figure is from Ref. [45].

particles [48–50]. This is due to the difference in final state interaction in these two cases: if the faster particle of a pair approaches and passes the other one from behind, the pair experiences a stronger final-state interaction than in the case when the faster particle has started in front of the slower one. These two situations are depicted in Fig. 14.20. Pion–kaon, and also

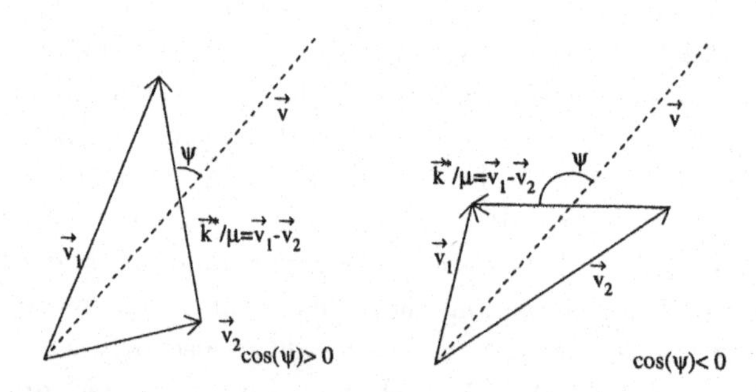

Fig. 14.20 Momentum vector diagrams showing the two selections of pairs of non-identical particles used for the determination of their emission order (from Ref. [49]).

pion–proton and kaon–proton correlations have been studied along these lines by the STAR Collaboration at RHIC [51, 52]. Figure 14.21 shows, as examples, the correlation functions for pion–kaon pairs of the same sign of charge (left panel), and of opposite charges (right panel). Dividing particle pairs in two groups according to Fig. 14.20, and forming the ratios of the corresponding correlation functions, some emission asymmetries between

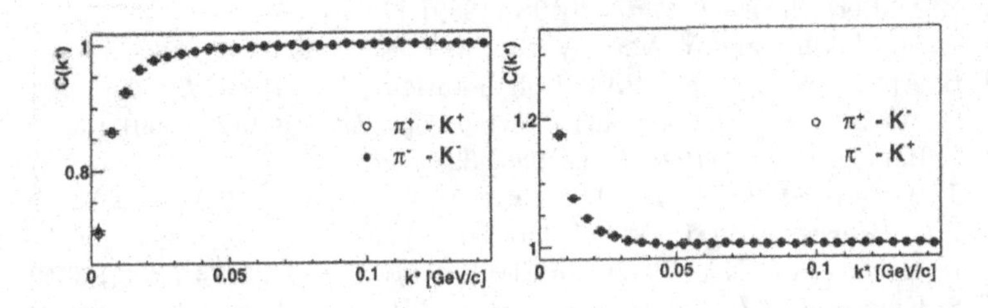

Fig. 14.21 Correlation functions for pion–kaon pairs with like (left panel) and unlike (right panel) charges from central Au+Au collisions at $\sqrt{s_{NN}} = 130$ GeV (from Ref. [52]).

various particles have been found. They show consistency with a model assuming transverse collective expansion (*blast wave model*). This provides an independent evidence for the radial flow resulting from a transverse radial expansion of the system.

References

[1] G. Goldhaber *et al.*, *Phys. Rev. Lett.* **3** (1959) 181.
[2] G. Goldhaber, S. Goldhaber, W. Lee and A. Pais, *Phys. Rev.* **120** (1960) 300.
[3] J. Bartke *et al.*, *Phys. Lett. B* **24** (1967) 163.
[4] K. Eskreys, *Acta Phys. Polon.* **36** (1969) 237.
[5] A. M. Cooper *et al.*, *Nucl. Phys. B* **139** (1978) 45.
[6] T. Akesson *et al.*, *Phys. Lett. B* **155** (1985) 128.
[7] R. Hanbury-Brown and R. Q. Twiss, *Philos. Mag.* **45** (1954) 663.
[8] G. I. Kopylov and M. I. Podgoretsky, *Sov. Phys. JETP* **42** (1975) 211.
[9] R. Lednicky, *Brazilian J. Phys.*, in press.
[10] M. Deutschmann *et al.*, *Nucl. Phys. B* **204** (1982) 333.
[11] G. I. Kopylov and M. I. Podgoretsky, *Yad. Fiz.* **15** (1972) 392; **18** (1973) 656; **19** (1974) 434 [*Sov. J. Nucl. Phys.* **15** (1972) 219; **18** (1973) 336; **19** (1974) 215]; G. I. Kopylov, *Phys. Lett. B* **50** (1974) 412.
[12] G. Cocconi, *Phys. Lett. B* **49** (1974) 459.
[13] F. B. Yano and S. E. Koonin, *Phys. Lett. B* **78** (1978) 556.
[14] M. Gyulassy, S. K. Kaufmann and L. W. Wilson, *Phys. Rev. C* **20** (1979) 2267.
[15] R. Weiner, *Phys. Lett. B* **232** (1989) 278; *ibid.* **242** (1990) 547.
[16] M. I. Podgoretskii, *Fiz. Elem. Chastits At. Yadra* **20** (1989) 628 [*Sov. J. Part. Nucl.* **20** (1989) 266].
[17] S. Pratt, *Phys. Rev. Lett.* **53** (1984) 1219.
[18] K. Kolehmainen and M. Gyulassy, *Phys. Lett. B* **180** (1986) 203.

[19] G. F. Bertsch, *Nucl. Phys. A* **498** (1989) 173c.
[20] A. N. Makhlin and Y. M. Sinyukov, *Z. Phys. C* **39** (1988) 69.
[21] H. Appelshäuser *et al.* (NA49 Collaboration), *Eur. Phys. J. C* **2** (1998) 661.
[22] G. Alexander, I. Cohen and E. Levin, *Phys. Lett. B* **452** (1999) 159.
[23] J. Bartke, *Phys. Lett. B* **174** (1986) 32.
[24] J. Barrette *et al.* (E877 Collaboration), *Nucl. Phys. A* **610** (1996) 227c.
[25] I. G. Bearden *et al.* (NA44 Collaboration), *Nucl. Phys. A* **610** (1996) 240c.
[26] H. Beker *et al.* (NA44 Collaboration), *Phys. Rev. Lett.* **74** (1995) 3340.
[27] S. Kniege *et al.* (NA49 Collaboration), *AIP Conf. Proc.* **828** (2006) 473.
[28] R. Stock, *Nucl. Phys. A* **544** (1992) 405c.
[29] NA35 and NA49 Collaborations, unpublished.
[30] Z. Chajecki (STAR Collaboration), *Nucl. Phys. A* **774** (2006) 599; and Debasish Das (STAR Collaboration), Poster No 21 at QM'06, to be published.
[31] I. G. Bearden *et al.* (NA44 Collaboration), *Phys. Lett. B* **455** (1999) 77.
[32] M. M. Aggarwal *et al.* (WA98 Collaboration), *Phys. Rev. Lett.* **85** (2000) 2895.
[33] R. Willson *et al.* (STAR Collaboration), *Nucl. Phys. A* **715** (2003) 619c.
[34] S. E. Koonin, *Phys. Lett. B* **70** (1977) 43.
[35] R. Lednicky and V. L. Lyuboschitz, *Yad. Fiz.* **35** (1982) 1316; *Sov. J. Nucl. Phys.* **35** (1982) 770.
[36] T. Siemiarczuk and P. Zieliński, *Phys. Lett. B* **24** (1967) 675.
[37] N. Akhababian *et al.*, *Z. Phys. C* **26** (1984) 245.
[38] G. N. Agakishiyev *et al.*, *Z. Phys. A* **327** (1987) 443.
[39] N. N. Ajitanand *et al.* (E895 Collaboration), *Proc. 15th Workshop Nuclear Dynamics*, Park City (1999).
[40] J. Barrette *et al.* (E814/E877 Collaboration), *Phys. Rev. C* **60** (1999) 054905.
[41] H. Boggild *et al.* (NA44 Collaboration), *Phys. Lett. B* **458** (1999) 181.
[42] H. Appelshäuser *et al.* (NA49 Collaboration), *Phys. Lett. B* **467** (1999) 21.
[43] H. P. Gos (STAR Collaboration), *Eur. Phys. J. C* **49** (2007) 75.
[44] S. V. Afanasiev *et al.* (NA49 Collaboration), *Nucl. Phys. A* **698** (2002) 104c.
[45] J. Adams *et al.* (STAR Collaboration), *Phys. Rev. C* **74** (2006) 064906.
[46] P. Chung *et al.* (E895 Collaboration), *Phys. Rev. Lett.* **91** (2003) 162301.
[47] C. Blume *et al.* (NA49 Collaboration), *Nucl. Phys. A* **715** (2003) 55.
[48] C. J. Gelderloos *et al.*, *Phys. Rev. Lett.* **75** (1995) 3082.
[49] R. Lednicky *et al.*, *Phys. Lett. B* **373** (1996) 30.
[50] R. Kotte *et al.* (FOPI Collaboration), *Eur. Phys. J. A* **6** (1999) 185.
[51] J. Adams *et al.* (STAR Collaboration), *Phys. Rev. Lett.* **91** (2003) 262302.
[52] A. Kisiel *et al.* (STAR Collaboration), *Acta Phys. Polon. B* **35** (2004) 47.

Chapter 15

Collective Flow

In collisions of high energy nuclei, in which a large number of secondary particles are produced, occurence of some multiparticle correlations, or *collective phenomena*, might be expected. A correlated emission of produced particles, called "flow", was observed already in nuclear collisions at low energies, and also shows up at relativistic energies. The term "flow" should be understood as a phenomenological description of a collective expansion, and, in fact, hydrodynamical models are quite successful in describing the observed features of the data.

There are three types of flow, as depicted in Fig. 15.1.

In strictly central collisions of spherical nuclei the expansion should be isotropic in the plane perpendicular to the collision axis (the "transverse" plane), as in this case there is no defined reaction plane. This is *radial flow.* Its effect is observed in the transverse spectra of the produced particles: the observed spectra are the combined result of thermal emission and radial flow. The appropriate description is the *blast wave model* which is discussed in detail in Chapter 10.

In a non-central collision of two nuclei with impact parameter $b > 0$ the collision axis (usually called the z-direction) and the impact parameter vector define the reaction plane, and some anisotropy of particle distribution in the transverse plane may occur. This is called the *anisotropic flow.* Figure 15.2 defines the coordinate system. The distribution of particles in the azimuthal angle ϕ in the transverse plane is usually analyzed in terms of the Fourier expansion [1]

$$E\frac{d^3N}{d^3p} = \frac{1}{2\pi}\frac{d^2N}{p_T\,dp_T\,dy}\left(1 + 2\sum_{n=1}^{\infty} v_n(p_T, y)\cos[n(\phi - \Psi_R)]\right) \tag{15.1}$$

where Ψ_R defines the position of the reaction plane (see Fig. 15.2). The method of estimating the reaction plane, and its resolution, was developed

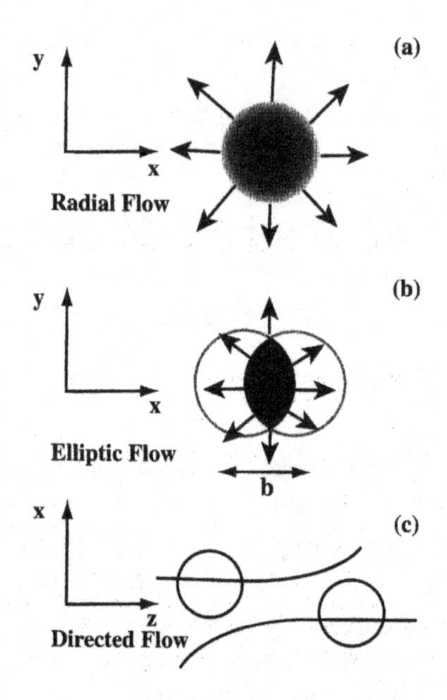

Fig. 15.1 Three types of flow phenomena.

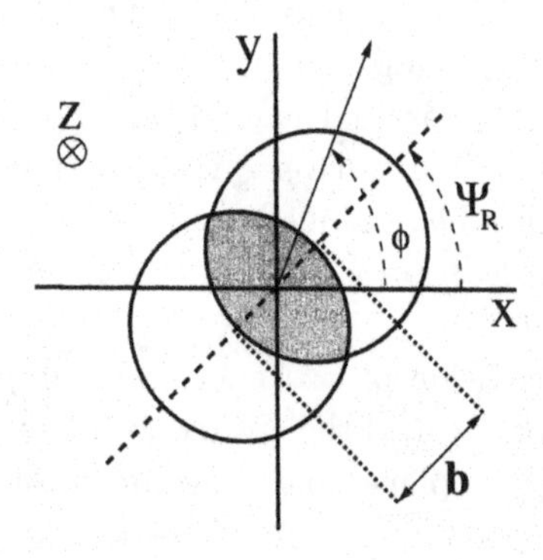

Fig. 15.2 The coordinate system and the reaction plane

in Ref. [2]. Only cosine terms appear in this Fourier series as sine terms vanish because of the reflection symmetry with respect to the reaction plane. The coefficients v_n can be calculated as the averaged values

$$v_n(p_T, y) = \langle \cos[n(\phi - \Psi_R)] \rangle \tag{15.2}$$

For an isotropic emission $v_n = 0$ for all n, non-zero values of v_n mean anisotropic flow. Odd Fourier harmonics have opposite sign in the forward and backward hemispheres, which is not the case for even harmonics. The most interesting is the second harmonic v_2, called the *elliptic flow*, as it characterizes the ellipticity of the azimuthal distribution of the produced particles. This is the dominant flow pattern. Higher even harmonics of the flow, mainly the v_4, have also been studied. They are much smaller than v_2 (for v_6 and v_8 only upper limits could have been given).

A noteworthy remark is that final state interactions should give rise to a positive value of v_2 [3]. This is due to various path lengths in the medium at different azimuthal angles, as for peripheral collisions the interaction region does not have azimuthal symmetry — see Fig. 15.2.

Figure 15.3 shows the elliptic flow v_2 of charged particles as a function of the energy of the collision. It increases with increasing energy from negative

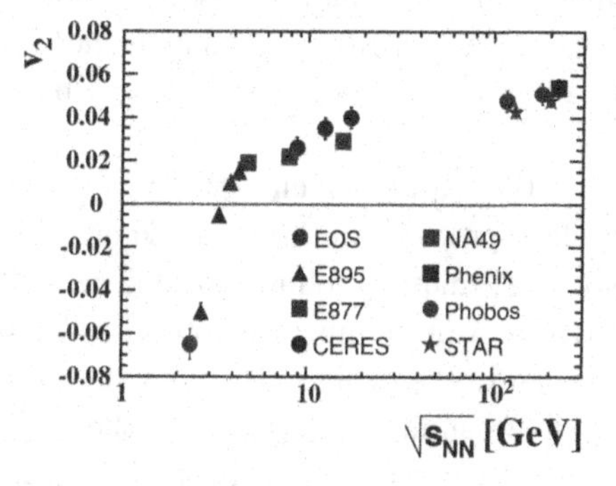

Fig. 15.3 Elliptic flow of charged particles as a function of the collision energy from AGS to the highest RHIC energies (from Ref. [4]).

values at low energies to positive ones at high energies. This means that at low energies there is an "out-of-plane" elliptic flow (*squeeze-out*) which then changes to "in-plane" elliptic flow. The transition takes place at about $\sqrt{s_{NN}} \cong 3$ GeV. At RHIC energies $v_2 \approx 0.15$–0.20.

Figure 15.4 shows the elliptic flow of charged particles at RHIC as a function of the collision centrality, characterized by the number of participating nucleons, N_{part}. As it can be expected, the largest elliptic flow is in the most peripheral collisions, and it decreases with the increasing number of participating nucleons (it should be zero for strictly central collisions).

Figure 15.5 shows the elliptic flow of charged particles for semi-central Au+Au collisions at four different energies, plotted as a function of pseu-

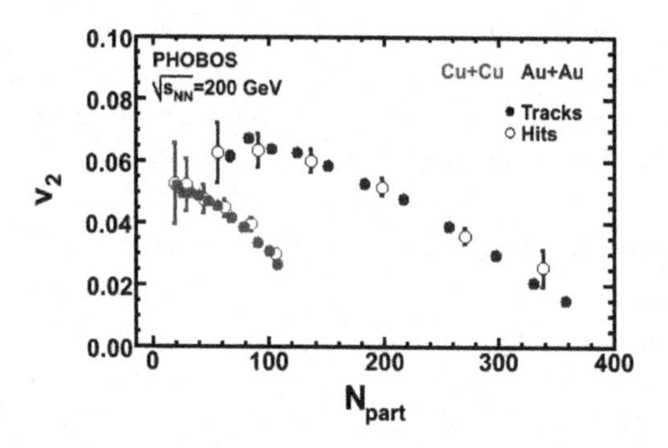

Fig. 15.4 Elliptic flow for Cu+Cu (lower points) and Au+Au (upper points) collisions at $\sqrt{s_{NN}} = 200$ GeV as a function of the number of participating nucleons (from Ref. [5]).

dorapidity. The elliptic flow is largest at midrapidity and falls off linearly towards larger values of $|\eta|$. This pattern seems to hold over a broad range of incident energies, however in Ref. [7] a flatter dependence of v_2 on rapidity in the central region has been reported for Pb+Pb collisions at the SPS.

Figure 15.6 shows the transverse energy dependence of the elliptic flow for identified particles: protons, antiprotons, kaons, pions and ϕ-mesons. Baryons/antibaryons and mesons follow two different curves. If, however, they are scaled by the relevant numbers of constituent quarks, n_q, (three for baryons/antibaryons and two for mesons), then they all fall on a universal curve. Figure 15.7 shows similar data from another experiment, which include also Λ and multi-strange hyperons, plotted as a function of transverse momentum. The n_q-scaling is evident. Recent measurement of the elliptic flow for "non-photonic" electrons from Au+Au collisions at $\sqrt{s_{NN}} = 200$ GeV, which are believed to result from decays of D-mesons, extends this scaling also to charmed quarks [10].

This is a very interesting observation, suggesting that thermalization takes place at quark level (not at hadronic level), and quark coalescence is the dominant mechanism of hadronization. In the framework of the quark coalescence hadronization dynamics [11], the elliptic flow for mesons, $v_{2,M}$, and for baryons, $v_{2,B}$, can be expressed in terms of the quark flow, $v_{2,q}$ as

$$v_{2,M}(p_T) \approx 2 \cdot v_{2,q}(p_T/2) \tag{15.3}$$

$$v_{2,B}(p_T) \approx 3 \cdot v_{2,q}(p_T/3) \tag{15.4}$$

what is confirmed by the data, and implies that the collective flow for quarks of all flavours is the same [12]. At RHIC energies $v_{2,q} \approx 0.08$.

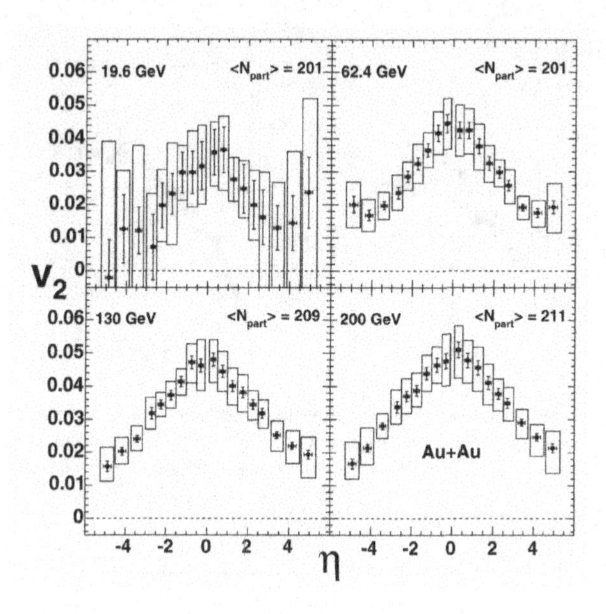

Fig. 15.5 Pseudorapidity dependence of elliptic flow of charged particles for the 40% most central Au+Au collisions at four beam energies. The values of $\sqrt{s_{NN}}$ and the average numbers of participating nucleons are given in each panel. Boxes indicate systematic uncertainties (from Ref. [6]).

The third type of flow is the *directed flow*. It is called so because it has a direction. The directed flow is characterized by the non-zero first harmonic $v_1 \neq 0$, and has opposite sign in the two hemispheres. It results mainly from momentum conservation: the flow of particles originating from one of the nuclei must be counterbalanced by the equal in magnitude, but opposite, flow of particles from the other nucleus. Figure 15.8 shows the directed flow of charged particles in Au+Au collisions at $\sqrt{s_{NN}} = 62$ GeV for different centralities, plotted as a function of pseudorapidity. As it could be expected, the directed flow is largest for most peripheral collisions (upper left corner of the figure), and decreases towards zero for the most central collisions (bottom right corner of the figure).

Results for identified particles reveal new and interesting features of this phenomenon: the directed flow for pions has the opposite sign as that of protons. This is shown in Fig. 15.9. The directed flow of protons increases linearly with transverse momentum, while a similar dependence for pions is very weak. Otherwise the directed flow for both types of particles decreases from peripheral towards central collisions — this is seen in Fig. 15.9.

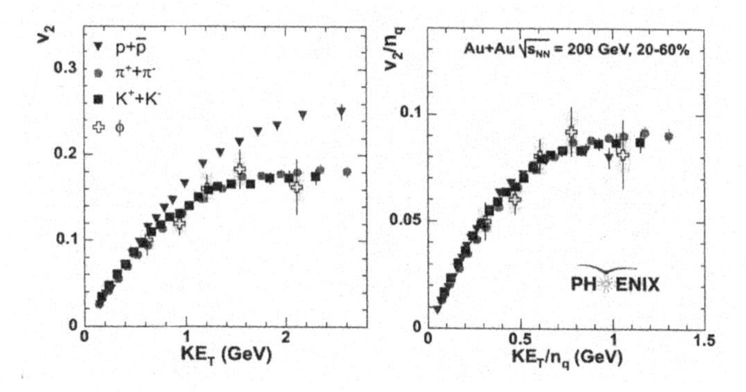

Fig. 15.6 Left panel: elliptic flow for identified particles (protons, antiprotons, kaons, pions and ϕ-mesons) in Au+Au collisions at $\sqrt{s_{NN}} = 200$ GeV as a function of transverse energy. Right panel: the same data scaled by the number of constituent quarks (from Ref. [8]).

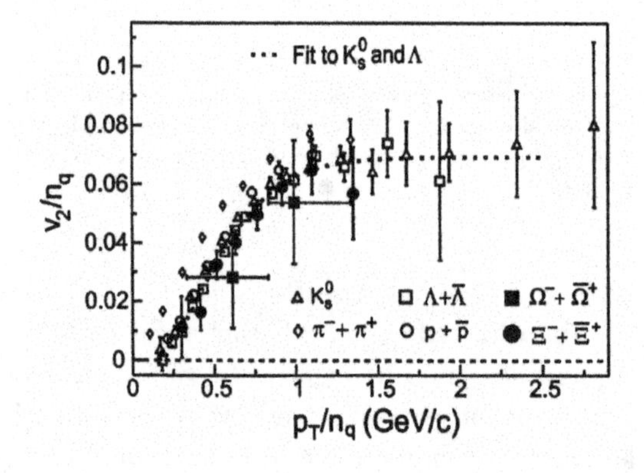

Fig. 15.7 Elliptic flow for different types of particles from Au+Au collisions at $\sqrt{s_{NN}} =$ 200 GeV, scaled by the number of constituent quarks, as a function of transverse momentum, similarly scaled. The dashed curve has been fitted to K_s^0 and Λ (from Ref. [9]).

Different behaviour of the directed flow for protons and pions, which was not observed at low energies, indicates that the proton flow and the pion flow have different physical origins. The directed flow of protons, as well as the elliptic flow, is believed to be sensitive to the equation of state of the hadronic (or quark) matter. Hydrodynamic description of both types of flow has, however, only a limited success: the dependence of v_2 on pseudorapidity (Fig. 15.5), and opposite signs of v_1 for protons and pions (Fig. 15.9), are difficult to explain within such approach.

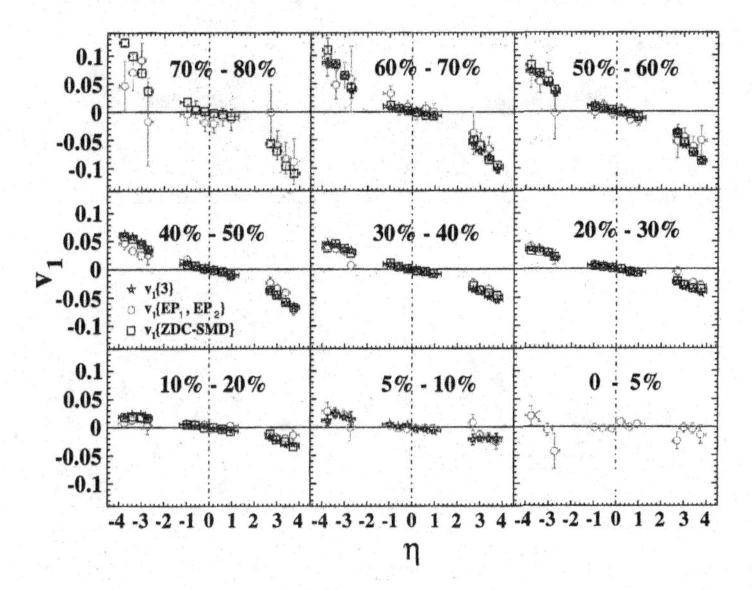

Fig. 15.8 Directed flow of charged particles from Au+Au collisions at $\sqrt{s_{NN}} = 62$ GeV for different centralities as a function of pseudorapidity. Different symbols indicate different methods of evaluation of the flow (from Ref. [13]).

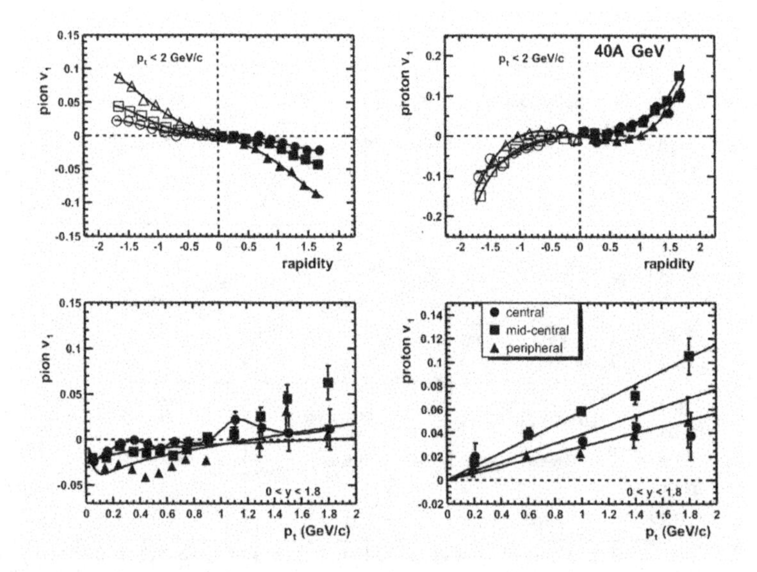

Fig. 15.9 Directed flow of pions (left) and protons (right) in Pb+Pb collisions at 40A GeV for different centralities, plotted as a function of rapidity (upper panels), and of transverse momentum (lower panels). The three centrality bins have been defined as follows: "central" — 12%, "mid-central" — 21%, and "peripheral" — 67% of the total hadronic inelastic cross section. The open points in the top graphs have been reflected about midrapidity (from Ref. [7]).

References

[1] S. Voloshin and Y. Zhang, *Z. Phys. C* **70** (1996) 665.
[2] A. M. Poskanzer and S. A. Voloshin, *Phys. Rev. C* **58** (1998) 1671.
[3] A. Capella and E. G. Ferreiro, *Phys. Rev. C* **75** (2007) 024905.
[4] B. Alessandro *et al.* (ALICE Collaboration), *J. Phys. G: Nucl. Part. Phys.* **32** (2006) 1295, and references therein.
[5] D. J. Hofman *et al.* (PHOBOS Collaboration), *J. Phys. G: Nucl. Part. Phys.* **34** (2007) S217.
[6] B. B. Back *et al.* (PHOBOS Collaboration), *Phys. Rev. Lett.* **94** (2005) 122303.
[7] C. Alt *et al.* (NA49 Collaboration), *Phys. Rev. C* **68** (2003) 034903.
[8] A. Taranenko *et al.* (PHENIX Collaboration), *J. Phys. G: Nucl. Part. Phys.* **34** (2007) S1069.
[9] J. Adams *et al.* (STAR Collaboration), *Phys. Rev. Lett.* **95** (2005) 122301.
[10] S. Sakai *et al.* (PHENIX Collaboration), *J. Phys. G: Nucl. Part. Phys.* **32** (2006) S551.
[11] D. Molnar and S. A. Voloshin, *Phys. Rev. Lett.* **91** (2003) 092301.
[12] J. Zimanyi, *Nucl. Phys. A* **774** (2006) 25.
[13] J. Adams *et al.* (STAR Collaboration), nucl-ex/0510053.

Chapter 16

Charmonium Suppression

Suppression of "charmonium"[a] in high energy nuclear collisions relative to elementary hadronic reactions was suggested in 1986 by Matsui and Satz [1] as a signal of a phase transition of nuclear matter to a deconfined state, or quark-gluon plasma. They argued that in a deconfined medium the strong colour field would "dissolve" charmonium by a mechanism similar to the Debye screening in solids. It is believed that charmonia have small radii, smaller or comparable to the colour field screening radius in the quark-gluon plasma. Then, if quark-gluon plasma is created in relativistic nuclear collisions, the relative yield of charmonia should be reduced as compared to elementary reactions. This would affect the J/ψ meson ($M = 3097$ MeV), and also higher states such as ψ' ($M = 3686$ MeV).[b] Figure 16.1 shows qualitatively the effect of charmonium suppression. Both charmonia: J/ψ and ψ', are visible as two peaks above the background which in this mass region comes from the so-called Drell–Yan process of dilepton production via virtual photons arising from hard collisions occuring between partons: quarks or gluons, $q + q(g) \to \gamma^* \to l^+ l^-$. The cross section for the Drell–Yan process can be calculated exactly within the perturbative QCD. This "elementary" cross section is small, and thus in nuclear collisions the Drell–Yan cross section should factorize as follows

$$\sigma_{\text{DY}}^{AB} = A \cdot B \cdot \sigma_{\text{DY}}^{\text{pp}} \tag{16.1}$$

where A and B are the mass numbers of the projectile and the target. This is verified by experimental data. Figure 16.2 shows the ratio of Drell–Yan cross sections measured in reactions involving various nuclei to the theoretically calculated cross sections for these processes, plotted against

[a] This term means the bound state of charmed quark and antiquark, $c\bar{c}$

[b] These are narrow isosinglet vector mesons which could be detected via decay into lepton pairs, $\mu^+\mu^-$ or e^+e^-.

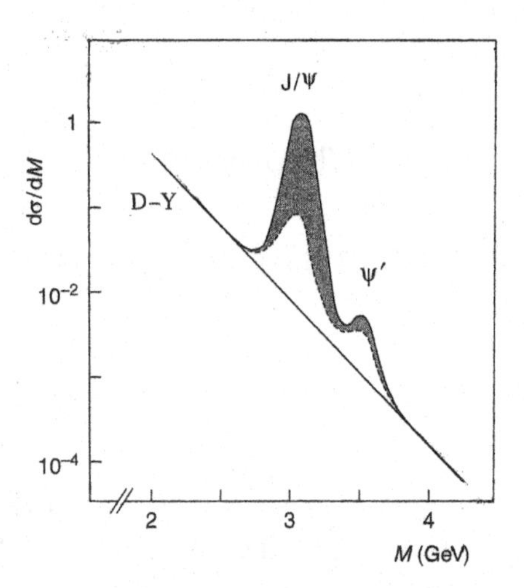

Fig. 16.1 Schematic visualisation of charmonium suppression in nuclear collisions relative to p+p reactions. The shaded area represents the amount of suppression (see text for details).

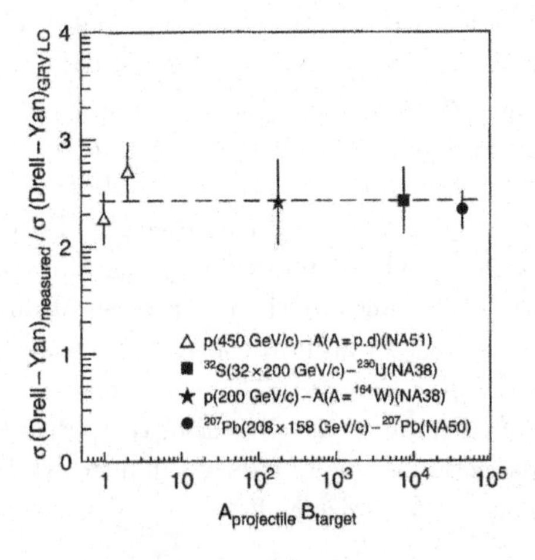

Fig. 16.2 Ratio of the measured Drell–Yan cross section to the calculated one plotted against the product of mass numbers of the colliding nuclei (from Ref. [2]).

the product of mass numbers of the colliding nuclei. This ratio remains constant all the way from p+p to Pb+Pb collisions. For lepton pairs arising from the Drell–Yan process the theory predicts a continuous mass spectrum falling as $d\sigma/dM \propto M^{-3}$, approximated with a straight line in Fig. 16.1. The charmonia peaks observed in elementary collisions, delineated with solid line in this Figure, in nuclear collisions should become smaller, as shown with dashed line. The difference between them, shown as the shaded areas, is the effect of suppression.

Charmonia can be most conveniently detected in their decay modes into muons, $\mu^+\mu^-$ (6% for J/ψ, 0.7% for ψ' [3]). As an example of experimental results, Figure 16.3 shows the measured effective mass distribution of muon pairs from Pb+Pb collisions at 158A GeV. The J/ψ peak is clearly seen, the ψ' appears rather as a "shoulder", but can also be analyzed. One should, however, remember that many J/ψ's are of secondary origin, in particular the ψ' is an important source of J/ψ's: $(56 \pm 1)\%$ of ψ' decay into "J/ψ + anything" [3].

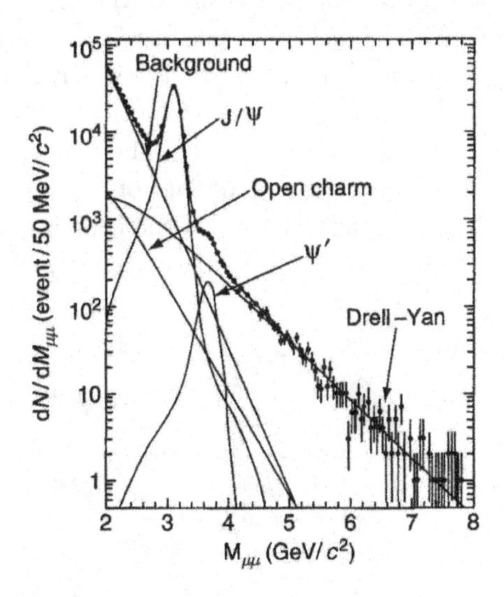

Fig. 16.3 Experimental mass spectrum of muon pairs from Pb+Pb collisions at 158A GeV. Different components of the total spectrum are shown (from Ref. [4]).

First experiments at the CERN SPS, using the dimuon magnetic spectrometer and oxygen and sulphur beams, showed indeed a suppression of J/ψ relative to proton–proton interactions, and this was initially taken as a confirmation of the above quoted theoretical prediction. Soon, however, it has been realized that in a dense medium the J/ψ meson should undergo some absorption, and this would lead to a certain depletion of the signal as compared to p+p interactions. The nuclear absorption cross section for charmonia have been measured in p+A reactions to be [5]:

$$\sigma_{\rm abs}(J/\psi) = (4.2 \pm 0.4)\text{mb} \tag{16.2}$$

$$\sigma_{\rm abs}(\psi') = (7.6 \pm 1.1)\text{mb} \tag{16.3}$$

As the Drell–Yan background is not the same in various reactions, the effect of J/ψ suppression is studied via the "double ratio": the J/ψ yield over the background in nuclear collisions divided by the J/ψ yield over the background in p+p reactions. A similar procedure is followed for ψ'. The so defined relative yield of charmonia is usually plotted as a function of the length, L, which the J/ψ has to traverse in matter - see Fig. 16.4. It determines the amount of nuclear absorption of J/ψ. L reaches its maximal value for central collisions. Another characteristic length is l, also defined in Fig. 16.4. l is the initial transverse size of the interaction region (the region of overlap of colliding nuclei). It also increases with centrality, and is related to the impact parameter b, $l = 2R - b$, reaching a maximal value of $2R$ for strictly central collisions. l is related to the number of participating nucleons, $N_{\rm part}$, for small l $N_{\rm part} = \text{const}\cdot l^3$. The two lengths, L and l, are interrelated and can be computed as functions of the impact parameter if a spherical shape and some density profile of the colliding nuclei is assumed (*e.g.* the Saxon–Woods distribution). Experimental results on charmonium

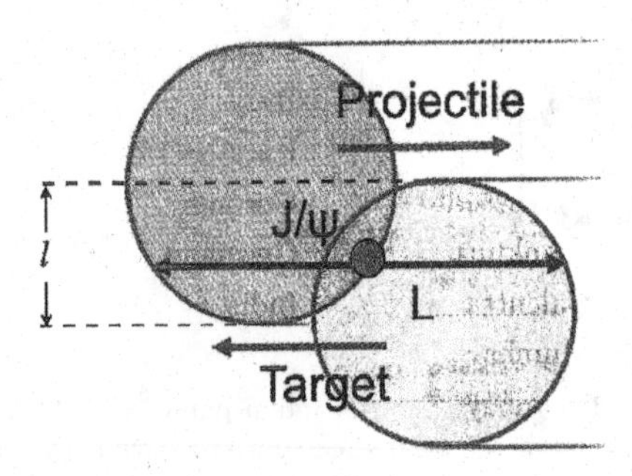

Fig. 16.4 Definition of the two characteristic lengths, L and l, in a collision of two nuclei with impact parameter b (from Ref. [6]).

suppression in nuclear collisions relative to p+p reactions are shown in Fig. 16.5 for J/ψ and in Fig. 16.6 for ψ'. The "double ratio" defined above is plotted against the length L. It can be seen that up to L = 6–7 fm for J/ψ, and up to L = 4–5 fm for ψ', the decrease of the yield of charmonia can be described by nuclear absorption (shown with straight lines with error bands), while a clear discontinuity occurs at larger values of L. This happens in semi-central and central Pb+Pb collisions, and recently has also been seen in central In+In collisions [7]. Discontinuity in the charmonium suppression pattern is emphasized if the relative charmonium yield is normalized to the nuclear absorption cross section. This is shown for J/ψ in Fig. 16.7.

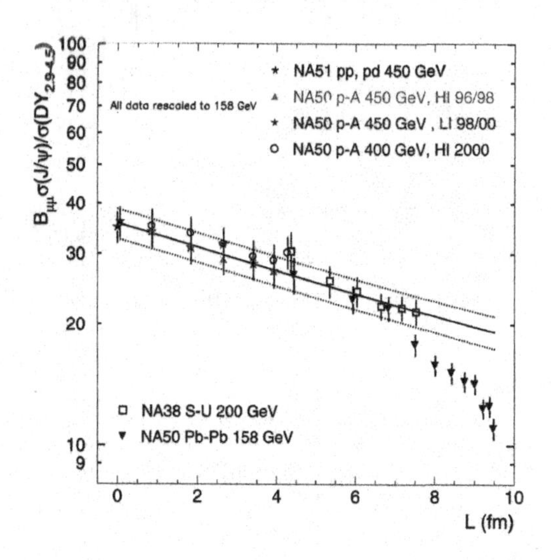

Fig. 16.5 The J/ψ yield relative to the Drell-Yan background plotted as a function of L (from Ref. [5]).

The nature of this abnormal suppression remains so far without a satisfactory explanation. The proposed mechanism of this being due to charmonium interaction with nuclear "comovers" can give some additional suppression, but not a discontinuous behaviour observed in the data.

It is not clear whether the charmonium suppression observed at SPS energies will remain the same at higher energies. On the contrary, there are predictions that at higher energies the J/ψ suppression due to Debye screening may be overcome by the production of charmonia by recombination of primordially produced c and $\bar{c}$ quarks [8]. Preliminary results from RHIC for Au+Au collisions at $\sqrt{s_{NN}}$ = 200A GeV show, however, a J/ψ suppression pattern similar to that observed at SPS energies [9].

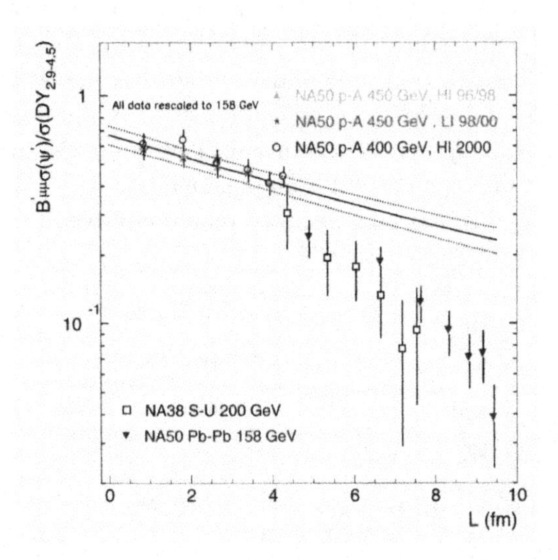

Fig. 16.6 The ψ' yield relative to the Drell-Yan background plotted as a function of L (from Ref. [5]).

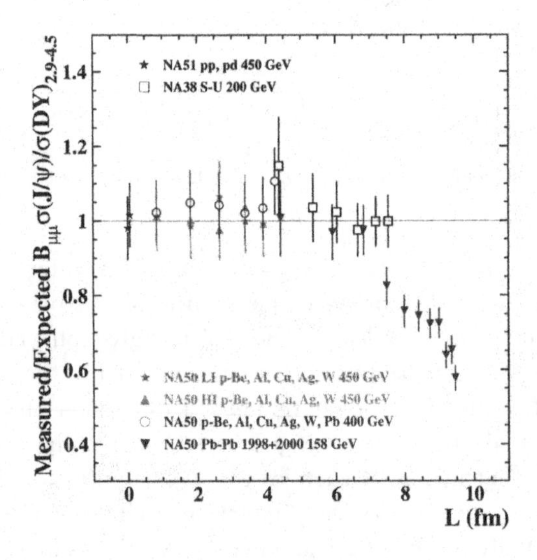

Fig. 16.7 Measured J/ψ production normalized by the ordinary nuclear absorption, plotted as a function of L (from Ref. [5]).

References

[1] T. Matsui and H. Satz, *Phys. Lett. B* **178** (1986) 416.
[2] M. C. Abreu *et al.* (NA50 Collaboration), *Nucl. Phys. A* **610** (1996) 404c.
[3] W. M. Yao *et al.* (Particle Data Group), *J. Phys. G: Nucl. Part. Phys.* **33** (2006) 1.
[4] M. C. Abreu *et al.* (NA50 Collaboration), *Phys. Lett. B* **450** (1999) 456.
[5] L. Ramello (NA50 Collaboration), *Nucl. Phys. A* **774** (2006) 59.
[6] I. Maiani, *Nucl. Phys. A* **774** (2006) 14.
[7] R. Arnaldi *et al.* (NA60 Collaboration), *Nucl. Phys. A* **774** (2006) 711.
[8] P. Braun-Munzinger and J. Stachel, *Nucl. Phys. A* **690** (2001) 119.
[9] H. Pereira da Costa (PHENIX Collaboration), *Nucl. Phys. A* **774** (2006) 747.

Chapter 17

Puzzle in Di-Lepton Mass Spectrum

Invariant mass spectra of lepton pairs (e^+e^- or $\mu^+\mu^-$) should reveal peaks corresponding to direct decays of various mesons ($\rho, \omega, \phi, J/\psi$) into a lepton pair, over a continuous background resulting from three-body decays, mainly the so-called *Dalitz decays* of π^0, η, η', and ω into a lepton pair and a photon, but also from other many-body decays, including those of charmed hadrons.

Di-electron mass spectra up to about 1.5 GeV for various colliding systems have been studied by the CERES experiment at the CERN SPS, and have shown some unexpected features. Figure 17.1 shows the e^+e^- mass spectrum from p+Be collisions at 450 GeV. Let us note that the spectrum spans over five orders of magnitude. Contributions to this spectrum from various decay channels are shown with thin lines, and their sum with a thick line. It can be seen that this sum, called by the authors *the hadronic cocktail*, describes quite well the experimental data within statistical and systematical uncertainties, shown in Fig. 17.1 correspondingly with error bars and a grey band. In the mass spectrum below 200 MeV the spectrum is overwhelmingly π^0 Dalitz decay, above 200 MeV heavier mesons contribute. The e^+e^- invariant mass spectrum for p+Au collisions at the same energy is very similar to that for p+Be, and can also be described by the "hadronic cocktail" [1].

For collisions of heavy nuclei the situation turns out, however, to be quite different. Figure 17.2 shows the e^+e^- invariant mass distribution for Pb+Au collisions at 158A GeV, also measured by the CERES Collaboration [2]. The "hadronic cocktail" calculated in the same way as previously does not describe the data, in the mass range between the π and the ρ a significant excess of electron pairs over the "cocktail" is observed.

The $\mu^+\mu^-$ invariant mass spectrum in In+In collisions at 158A GeV in

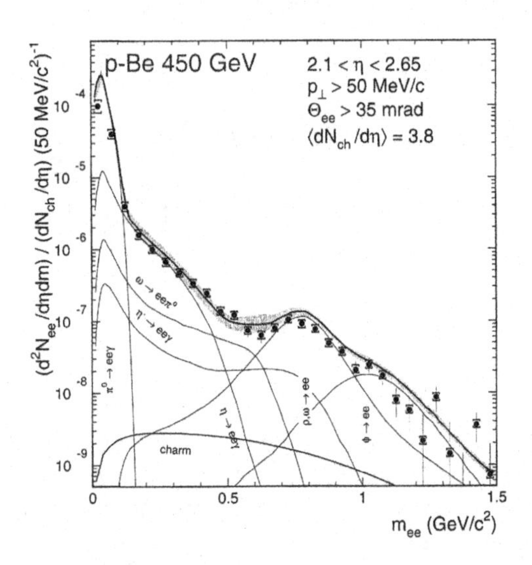

Fig. 17.1 Invariant mass spectrum of e^+e^- pairs from p+Be collisions at 450 GeV (from Ref. [1]). In this, and in subsequent figures, the *hadronic cocktail* of Dalitz- and direct meson decays is shown separately (thin lines), and summed (thick line).

the same mass range was studied by the NA60 experiment at CERN [3]. A similar excess below the ρ mass has been observed. Figure 17.3 shows, in the left panel, the results before and after background subtraction, and, in the right panel, only the "excess" and resonance peaks.

Theoretical ideas aiming at an explanation of the observed excess in dilepton mass spectrum below the ρ mass, assume a modification of the spectral shape of vector mesons (mainly of the ρ meson) in a dense medium. According to Brown and Rho [4], meson masses should decrease linearly as a function of the medium density: $m/m_0 \cong 1 - k(\rho/\rho_0)$, where ρ is the density of the medium, ρ_0 is the standard nuclear density, and k is a free parameter called *the shift parameter.* This relation is called *Brown–Rho scaling.* On the other hand, according to Rapp and Wambach [5], the effect of a dense medium should be broadening of mesons: $\Gamma/\Gamma_0 \cong 1 + k_1(\rho/\rho_0)$, where Γ is the in-medium meson width, Γ_0 is its standard width, and k_1 is a free parameter. In principle, one could also think of simultaneous occurence of both effects: mass shift and resonance broadening. Experimental results, especially the high resolution data of NA60, disfavour the mass shift hypothesis, indicating only some broadening of resonance peaks. One should, however, keep in mind that the apparent width of a resonance peak is a

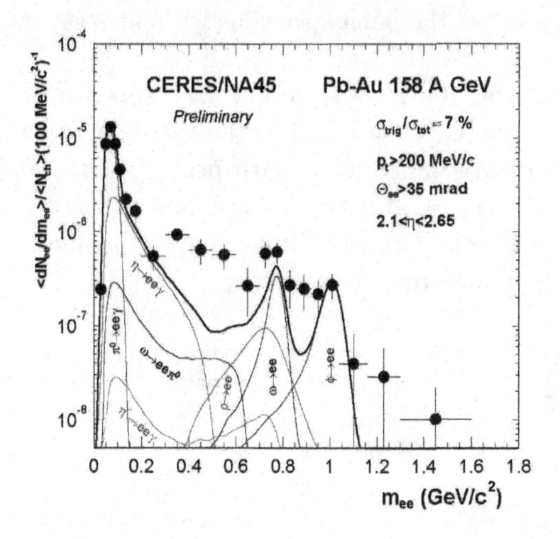

Fig. 17.2 Invariant mass spectrum of e^+e^- pairs from Pb+Au collisions at 158*A* GeV (from Ref. [2]).

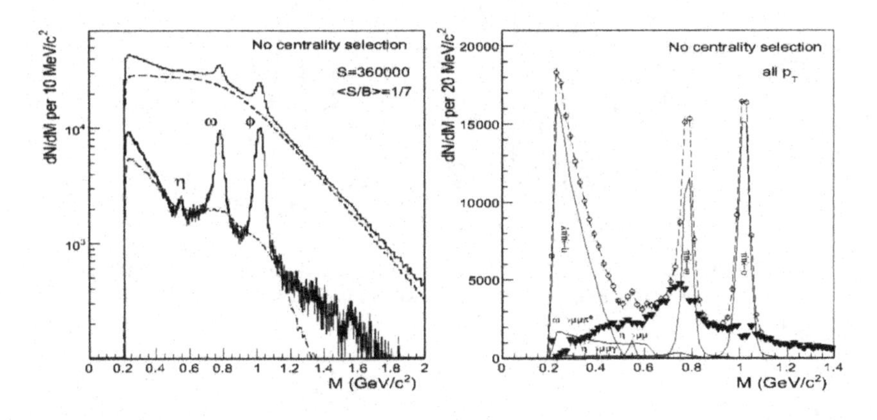

Fig. 17.3 Left panel: Invariant mass spectrum of $\mu^+\mu^-$ pairs from In+In collisions at 158*A* GeV (upper histogram), and the same spectrum after background subtraction (lower histogram). Right panel: total data (open circles) and isolated excess (full triangles) (from Ref. [3]).

convolution of the intrinsic width and of experimental resolution, and the latter is often difficult to estimate. An enhanced production of charmed mesons has been also considered as a possible source of the observed effect, but there is no independent evidence for it. Thus, the observed excess in dilepton spectra below the ρ mass remains without a satisfactory explanation.

Figure 17.4 shows the e^+e^- invariant mass spectrum in Au+Au collisions at $\sqrt{s_{NN}} = 200$ GeV measured by the PHENIX experiment at RHIC in a much wider mass range [6]. Sharp peaks of ω, ϕ, and J/ψ mesons are clearly seen. The region between the ϕ and the J/ψ is dominated by semi-leptonic decays of charmed hadrons, and can be understood. But the excess below the ρ mass remains a puzzle.

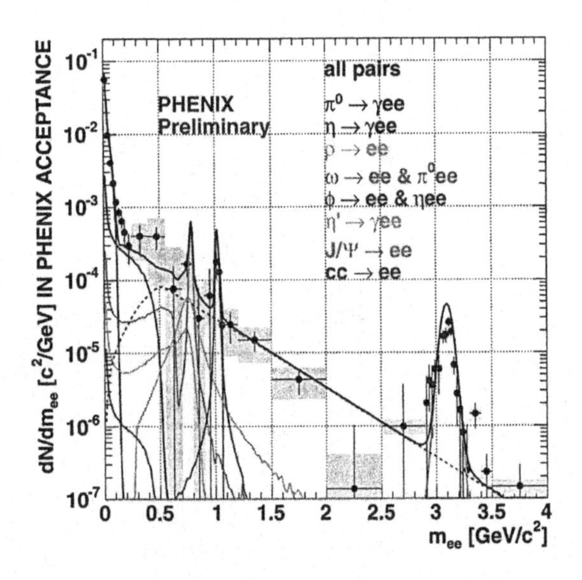

Fig. 17.4 Invariant mass spectrum of e^+e^- pairs after subtraction of combinatorial background, measured in minimum bias Au+Au collisions at $\sqrt{s_{NN}} = 200$ GeV (from Ref. [6]).

References

[1] G. Agakichiev *et al.*, *Eur. Phys. J. C* **4** (1998) 231.
[2] A. Cherlin and S. Yurevich (CERES Collaboration), *J. Phys. G: Nucl. Part. Phys.* **30** (2004) S1007.
[3] J. Seixas *et al.* (NA60 Collaboration), *J. Phys. G: Nucl. Part. Phys.* **34** (2007) S1023.
[4] G. E. Brown and M. Rho, *Phys. Rev. Lett.* **66** (1991) 2720.
[5] R. Rapp and J. Wambach, *Adv. Nucl. Phys.* **25** (2000) 1.
[6] S. Campbell (PHENIX Collaboration), *J. Phys. G: Nucl. Part. Phys.* **34** (2007) S1055.

Chapter 18

Direct Photons

In collisions of relativistic nuclei, the appearance of *direct* photons, *i.e.* photons which do not come from decays of final state hadrons, can be expected as resulting from interactions of various charged particles created in the collision, either at the partonic or at the hadronic level. Direct photons are considered to be an excellent probe of the early stage of the collision. This is because their mean free path length is very large as compared to the size of the system formed after the collision. Thus photons created at the early stage leave the system without suffering any interaction, and retain information about this stage, in particular about its temperature.[a] High transverse momentum photons can provide information whether the observed suppression of high p_T particles is an initial- or a final-state effect.

Direct photons are single photons. Depending on their origin, one distinguishes two kinds of them: *thermal* and *prompt*. Thermal photons can be emitted from the quark-gluon plasma, or from the hot hadronic matter (the hadronic gas). Thermal photons from quark-gluon plasma come from various electromagnetic processes involving quarks,[b] and from strong processes involving quarks and gluons: Compton scattering $q(\bar{q})\,g \rightarrow q(\bar{q})\,\gamma$, annihilation $q\,\bar{q} \rightarrow g\,\gamma$, and also from reactions $g\,q \rightarrow g\,q\,\gamma$, $q\,q \rightarrow q\,q\,\gamma$, $q\,q\,\bar{q} \rightarrow q\,\gamma$, $g\,q\,\bar{q} \rightarrow g\,\gamma$. The complete calculation of the photon emission rate from quark-gluon plasma to order α_s is given in Ref. [1]. Thermal photons from the hadronic gas may come from the processes $\pi\,\pi \rightarrow \rho\,\gamma$, $\pi\,\rho \rightarrow \pi\,\gamma$, $\pi\,\pi \rightarrow \eta\,\gamma$, $\pi\,\eta \rightarrow \pi\,\gamma$, single photons also result from decays $\rho \rightarrow \pi\,\pi\,\gamma$, $\omega \rightarrow \pi\,\gamma$. Calculation of the photon yield from hadronic reactions can be found in Ref. [2]. Thermal photons can be observed in the low

---

[a] Let us recall that hadronic spectra yield the temperature existing at the final stage of the collision process — see Chapter 10.

[b] Quarks carry electric charge $\frac{2}{3}e$ or $\frac{1}{3}e$.

p_T region. Prompt photons are believed to come from "hard" collisions of initial state partons belonging to the colliding nuclei. They will dominate the high p_T region. The yield of photons from "hard" collisions can be calculated using the perturbative QCD approach [3].

Direct photon extraction is an extremely difficult experimental task because their contribution amounts to only a few percent of the total yield, and rapidly decreases towards small p_T. The dominating photon source are final state hadron decays $\pi^0 \to 2\gamma$, $\eta \to 2\gamma$, *etc.* A convenient parameter to quantify the studied effect is

$$R = \frac{\gamma_{\text{inclusive}}}{\gamma_{\text{decay}}} = 1 + \frac{\gamma_{\text{direct}}}{\gamma_{\text{decay}}}$$

Attempts to extract the direct photon signal in 200A O+Au and S+Au collisions at the SPS yielded only upper limits [4]. In 158A GeV Pb+Pb collisions a significant signal of direct photons was observed for $p_T \geq 1.5$ GeV/c, and an upper limit was set for $0.5 \leq p_T < 1.5$ GeV/c [5]. Figure 18.1 shows these results, together with theoretical predictions [6]. The sum of contributions from quark-gluon plasma, hadronic gas, and perturbative QCD "hard" processes describes well the experimental data. Two points

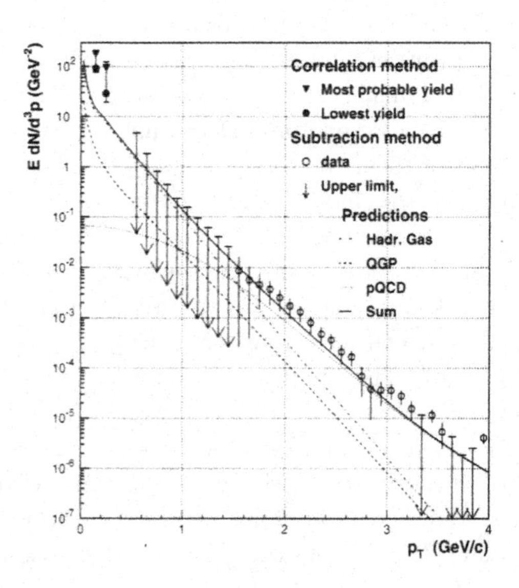

Fig. 18.1 Yield of direct photons from 158A GeV Pb+Pb collisions. Two points at low p_T have been obtained by the correlation method, all other points by the statistical subtraction method (from Ref. [7]).

at low p_T ($100 < p_T < 300$ MeV/c) which appear in this plot, have been added later. They result from a pioneering interferometry analysis of direct photons [7], while the yield of direct photons at higher p_T was obtained using the statistical subtraction method. These points, and even their lower limits obtained from the correlation strength parameter, appear to lie above the theoretical predictions.

A stronger signal of direct photons has been recently measured at RHIC in p+p, d+Au, and Au+Au collisions by the PHENIX [8] and STAR [9] collaborations. Apart from the standard statistical subtraction method, the PHENIX collaboration used other methods of single photon identification: tagging and external conversion $\gamma \rightarrow e^+ e^-$. Figures 18.2 and 18.3 show the yields of direct photons from d+Au and Au+Au collisions at $\sqrt{s_{NN}} =$ 200 GeV, plotted as functions of p_T, together with theoretical predictions (curves). Figures 18.4 and 18.5 show these yields displayed in terms of the

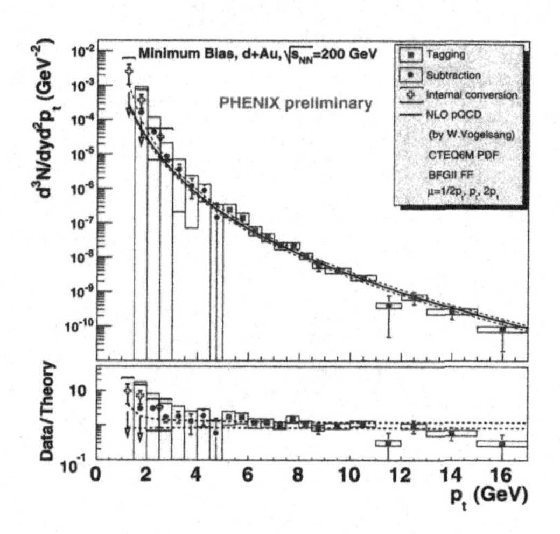

Fig. 18.2 Yield of direct photons from d+Au collisions at $\sqrt{s_{NN}} = 200$ GeV. Curves show theoretical predictions [10], a detailed comparison between data and theory is shown in the lower panel (from Ref. [11]).

parameter R. Let us note that the low p_T domain ($p_T < 1$ GeV/c) remains in these experiments not accessible.

As it has been stated at the beginning of this chapter, direct photons carry information about the temperature of the initial state formed in the collision. The appropriate analysis of experimental data for Pb+Pb collisions at the SPS ($\sqrt{s_{NN}} = 17.3$ GeV) yields the initial temperature

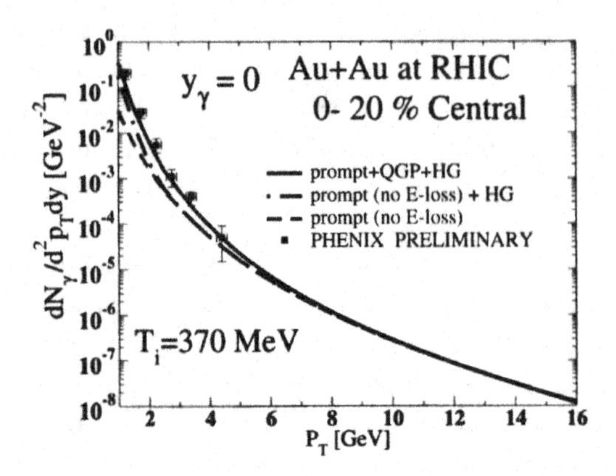

Fig. 18.3 Yield of direct photons from Au+Au collisions at $\sqrt{s_{NN}} = 200$ GeV. Curves show theoretical predictions, the full curve contains all components of the calculation (from Ref. [12]).

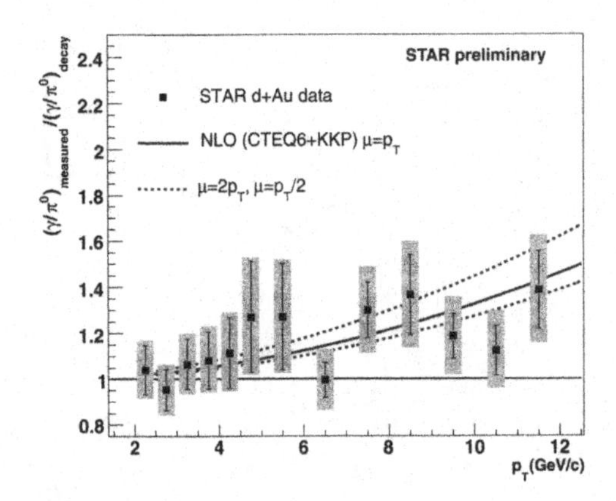

Fig. 18.4 Direct photon yield from d+Au collisions at $\sqrt{s_{NN}} = 200$ GeV plotted in terms of the ratio R. The full curve is the pQCD calculation, shown with its uncertainty limits (from Ref. [9]).

$T_i \cong 200$ MeV [13], while for Au+Au collisions at RHIC ($\sqrt{s_{NN}} = 200$ GeV) the initial temperature $T_i \cong 300$–400 MeV has been obtained[c] [14]. This is much higher than the freeze-out temperature $T_f \cong 120$ MeV obtained from the p_T spectra of pions and kaons (see Chapter 10). Thus direct photons provide us with a very interesting information, sheding light

[c] A deconfined state of quarks and gluons was assumed in this analysis.

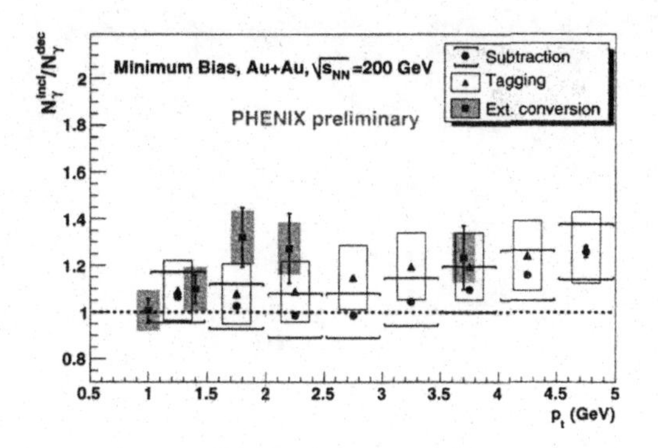

Fig. 18.5 Direct photon yield from Au+Au collisions at $\sqrt{s_{NN}} = 200$ GeV plotted in terms of the ratio R. Three types of points result from three different methods of extracting the direct photon signal (from Ref. [11]).

on the evolution of the collision process.

References

[1] P. Arnold *et al.*, *J. High Energy Phys.* **06** (2002) 030.
[2] J. Alam *et al.*, *Phys. Rev. C* **71** (2005) 059802.
[3] D. M. Zhou *et al.*, *Phys. Lett. B* **638** (2006) 461.
[4] R. Albrecht *et al.* (WA80 Collaboration), *Phys. Rev. Lett.* **76** (1976) 3506.
[5] M. M. Aggarwal *et al.* (WA98 Collaboration), *Phys. Rev. Lett.* **85** (2000) 3595.
[6] S. Turbido *et al.*, *Phys. Rev. C* **69** (2004) 014903.
[7] D. Peressounko *et al.* (WA98 Collaboration), *J. Phys. G: Nucl. Part. Phys.* **30** (2004) S1065.
[8] S. S. Adler *et al.* (PHENIX Collaboration), *Phys. Rev. D* **71** (2003) 071102(R); *Phys. Rev. Lett.* **94** (2005) 232301.
[9] M. J. Russcher *et al.* (STAR Collaboration), *J. Phys. G: Nucl. Part. Phys.* **34** (2007) S1033.
[10] L. E. Gordon and W. Vogelsang, *Phys. Rev. D* **48** (1993) 3136.
[11] D. Peressounko *et al.* (PHENIX Collaboration), *J. Phys. G: Nucl. Part. Phys.* **34** (2007) S869.
[12] C. Gale, *Nucl. Phys. A* **774** (2006) 335.
[13] J. Alam *et al.*, *Phys. Rev. C* **63** (2001) 021901.
[14] J. Alam, *J. Phys. G: Nucl. Part. Phys.* **34** (2007) S865.

Chapter 19

High Transverse Momenta

Particles emitted with high transverse momenta ($p_T \geq$ 2–3 GeV/c) are believed to result from "hard" scattering processes. As the distance characteristic for hard scattering is small, of the order of $1/p_T \leq 0.1$ fm, such processes are "point-like", and thus the yield of high p_T particles from collisions involving nuclei should scale with the number of elementary binary collisions, $N_{\rm coll}$. Medium effects can modify this scaling. This can be conveniently studied by a measurement of the so-called *nuclear modification factor* $R_{\rm AA}$. This quantity is defined as the ratio of the particle yield in A+A collisions, normalized to the number of binary collisions, to the corresponding yield in p+p collisions

$$R_{\rm AA}(p_T, \eta) = \frac{1}{\langle N_{\rm coll}^{\rm AA} \rangle} \frac{d^2N^{\rm AA}/dp_T\, d\eta}{d^2N^{\rm pp}/dp_T\, d\eta} \tag{19.1}$$

where $N_{\rm coll}^{\rm AA}$ is the number of binary collisions in an A+A collision given by the Glauber model calculation. If there is no medium effects, the yields should scale with $N_{\rm coll}$ and $R_{AA} = 1$.

The nuclear modification factor as a function of transverse momentum has been measured at the SPS up to p_T = 4–5 GeV/c (NA49 [1], NA57 [2], and WA98 [3]), and at RHIC up to $p_T \approx 20$ GeV/c (PHENIX [4, 5], PHOBOS [6, 7], and STAR [8]).

At medium transverse momenta some enhancement in R_{AA} is observed. One should, however, keep in mind that for $p_T \leq$ 2–3 GeV/c "soft" processes dominate, and particle yields scale with the number of participants, $N_{\rm part}$, rather than with the number of binary collisions, $N_{\rm coll}$, and thus the normalization used in the definition of R_{AA} is not correct. At higher values of p_T $R_{AA} \approx 1$ for d+Au collisions, while for Au+Au collisions $R_{AA} < 1$, and decreases with increasing p_T.

Figure 19.1 shows the nuclear modification factor for neutral pions from

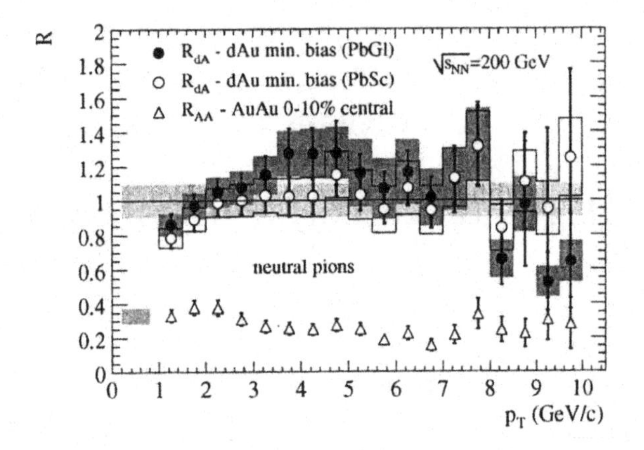

Fig. 19.1 Nuclear modification factor for neutral pions from d+Au collisions at $\sqrt{s_{NN}} =$ 200 GeV, plotted as a function of transverse momentum. Full and empty round points are the data obtained from two different calorimeters. The bands around the data points show systematic errors, while the shaded band around unity indicates the normalization uncertainty. The nuclear modification factor in the 10% most central Au+Au collisions is also shown for comparison (from Ref. [4]).

d+Au collisions at $\sqrt{s_{NN}} = 200$ GeV, plotted as a function of transverse momentum.

Figure 19.2 shows the nuclear modification factor for π^0 and η mesons from Au+Au collisions at $\sqrt{s_{NN}} = 200$ GeV, plotted as a function of transverse momentum. Results for direct photons are also given in this Figure. They show no suppression at high p_T, as it should be expected for non-interacting probes.

In Fig. 19.3 results from RHIC for π^0 mesons are shown again, together with those for charged hadrons, and, separately, for charmed and beauty particles. The suppression for charmed and beauty mesons has been measured indirectly, using non-photonic electrons which are expected to originate mainly from semileptonic decays of the corresponding heavy mesons (see Chapter 12). A measurement of R_{AA} for J/ψ mesons has been also reported [12]. All hadrons seem to show similar suppression pattern.

Figure 19.4 displays the nuclear modification factor as a function of centrality, quantified by the number of participants, for neutral pions with transverse momenta $2.0 < p_T < 2.5$ GeV/c from Pb+Pb collisions at $\sqrt{s_{NN}} = 17.3$ GeV.

Figure 19.5 shows similar dependence for both neutral pions and charged

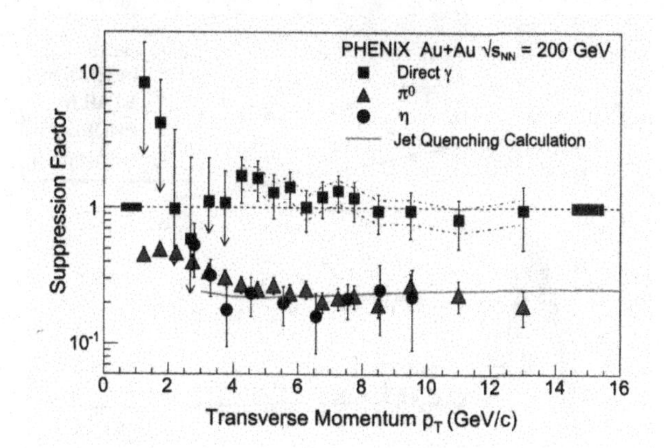

Fig. 19.2 Nuclear modification factor for π^0 and η mesons (triangles and circles), and for direct photons (squares), from Au+Au collisions at $\sqrt{s_{NN}} = 200$ GeV, plotted as a function of transverse momentum (from Ref. [9]).

hadrons with $p_T > 4.5$ GeV/c in Au+Au collisions at $\sqrt{s_{NN}} = 200$ GeV. In this p_T range all hadrons are suppressed, the suppression increases with increasing N_{part}.

It has also been found that the R_{AA} ratios for $(\pi^+ + \pi^-)$ mesons in Cu+Cu and Au+Au samples with similar numbers of binary collisions for their respective centralities show similar behaviour [13]. This shows that the energy loss depends on the geometry of the system.

As an alternative to R_{AA} the *central-to-peripheral ratio* R_{CP} is being used. It is defined as the ratio of the differential yields in central collisions to those in peripheral ones, both normalized to the corresponding numbers of binary collisions

$$R_{CP}(p_T, y) = \frac{\langle N_{\text{coll}}^{\text{Peripheral}} \rangle}{\langle N_{\text{coll}}^{\text{Central}} \rangle} \frac{d^2 N^{\text{Central}}(p_T, y)\, dp_T\, dy}{d^2 N^{\text{Peripheral}}(p_T, y)\, dp_T\, dy} \tag{19.2}$$

Again, in absence of medium effects both yields should scale with the number of binary collisions, N_{coll}, and $R_{CP} = 1$.

The BRAHMS Collaboration studied the dependence of R_{CP} for charged hadrons in d+Au collisions at $\sqrt{s_{NN}} = 200$ GeV on p_T and η [14]. Figure 19.6 shows the R_{CP} ratios for d+Au collisions at $\sqrt{s_{NN}} = 200$ GeV as a function of transverse momentum and pseudorapidity. No suppression is observed in the central region ($|\eta| \leq 1$), but a substantial change of R_{CP} as a function of η has been found: hadron yields are suppressed at forward rapidities. Also, the suppression is larger for more central collisions.

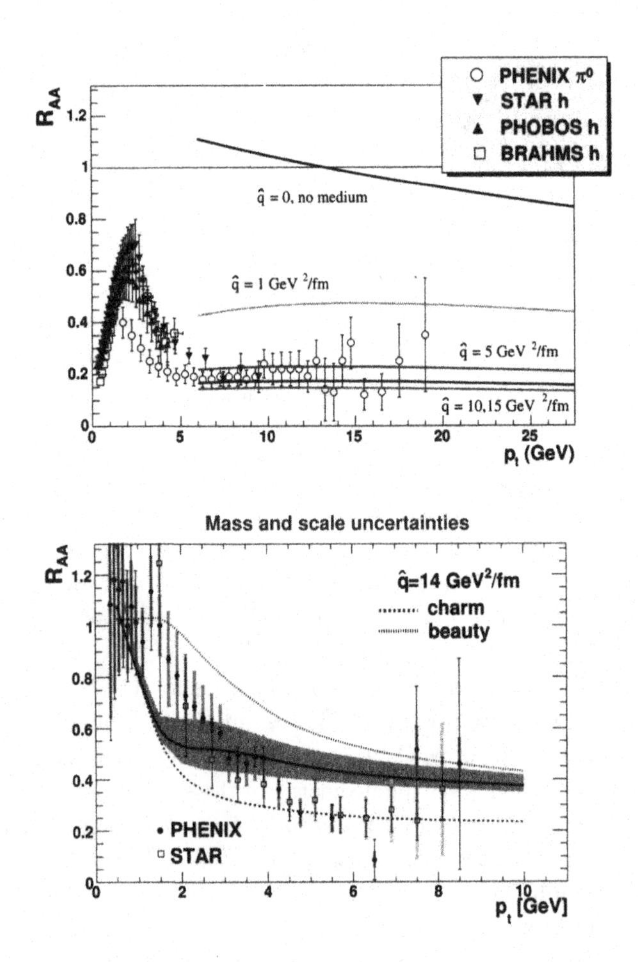

Fig. 19.3 Upper panel: nuclear modification factor for π^0 mesons and for light charged hadrons from Au+Au collisions at $\sqrt{s_{NN}}$ = 200 GeV, plotted as a function of transverse momentum. Data combined from all four experiments at RHIC. The curves have been calculated for different values of the in-medium transport parameter $\hat{q}$ (see text). The Figure is from Ref. [10]. Lower panel: similar data for charmed and beauty particles as obtained by PHENIX and STAR collaborations. Shaded bands indicate experimental uncertainties (from Ref. [11]).

All these observations, whether using R_{AA} or R_{CP}, indicate the presence of medium effects which set in for more central collisions. The suppression effect is strong: in central Au+Au collisions at $\sqrt{s_{NN}} = 200$ GeV, and at high transverse momenta ($p_T \geq 7$ GeV/c), the relative particle yields are

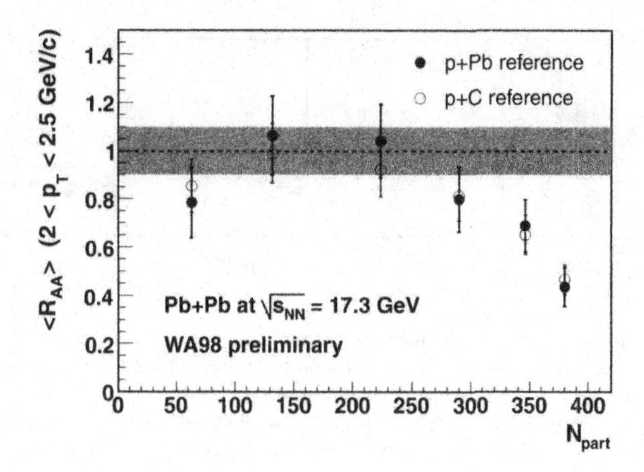

Fig. 19.4 Centrality dependence of the nuclear modification factor for π^0 mesons with $2.0 < p_T < 2.5$ GeV/c from Pb+Pb collisions at $\sqrt{s_{NN}} = 17.3$ GeV (from Ref. [3].)

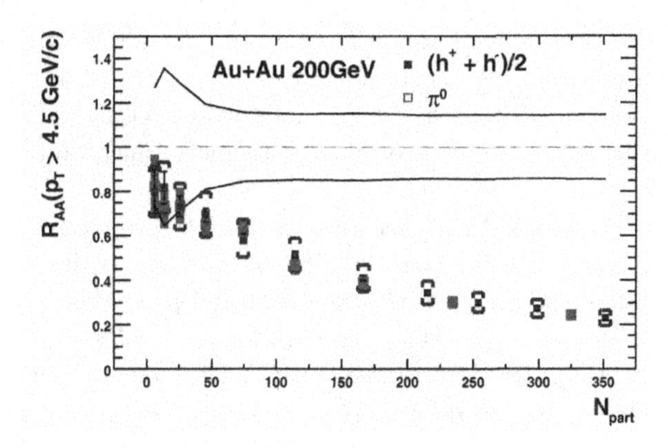

Fig. 19.5 Centrality dependence of the nuclear modification factor for neutral pions and charged hadrons with $p_T > 4.5$ GeV/c in Au+Au collisions at $\sqrt{s_{NN}} = 200$ GeV (from Ref. [5]). Bands around unity indicate the normalization uncertainty.

suppressed by a factor of 3–4 with respect to unity.

As it has already been stated, high transverse momentum particles are believed to result from "hard" scattering of partons. Hard scattering of quarks or gluons leads to the production of hadronic jets. Their attenuation in the medium is called *jet quenching.* The interaction in the medium of a high-p_T quark or gluon produced in an elementary hard collision is described as a multiple scattering process, and particle distributions are

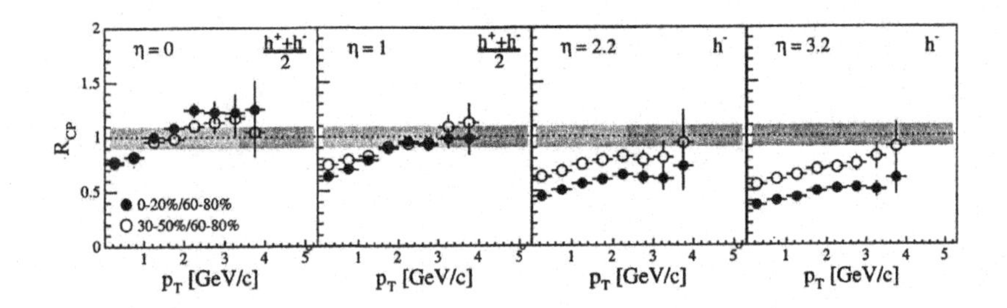

Fig. 19.6 The central-to-peripheral ratios for d+Au collisions at $\sqrt{s_{NN}} = 200$ GeV as a function of transverse moemntum and pseudorapidity. The filled points represent the central-to-peripheral (0–20% over 60–80%) ratio, and the open points the semicentral-to-peripheral (30–50% over 60–80%) ratio. The shaded band around unity indicates the uncertainty in the evaluation of the number of binary collisions. From Ref. [14].

assumed to follow a diffusion equation in transverse space. The coefficient in this equation, $\hat{q}$, is called the *jet transport parameter*, (or *transport coefficient*, or *quenching parameter*), and can be interpreted as the measure of the momentum broadening per unit length [15]. This is the main parameter to be determined from fits to experimental data. The fit to the light meson suppression, shown with curves in Fig. 19.3, leads to quite large values of the transport coefficient, but determined with a large uncertainty: $\hat{q} \simeq 5$–15 GeV^2/fm [10].

Another interesting characteristics of particle production with high transverse momenta are the ratios of different particle species. At the SPS energies it was already noticed that the K/π and p/π ratios increase with increasing transverse momentum, the data for $p_T \geq 2.5$ GeV/c had, however, rather large errors. This tendency was confirmed at RHIC energies.

Figure 19.7 shows the kaon to pion ratios as functions of p_T for central and peripheral Au+Au collisions at $\sqrt{s_{NN}} = 200$ GeV. Figure 19.8 shows the proton-to-pion and antiproton-to-pion ratios as functions of p_T for Au+Au collisions at $\sqrt{s_{NN}} = 200$ GeV with different centralities. All these ratios increase with p_T, and the increase is faster in central collisions than in peripheral ones. Antiproton-to-proton ratios are similar in central and peripheral collisions, and show almost flat p_T dependence. The value of this ratio is $\bar{\text{p}}/\text{p} \cong 0.73$ [9], in agreement with statistical model predictions.

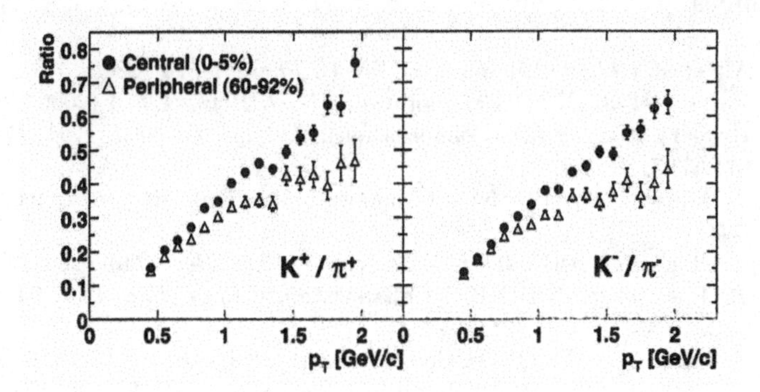

Fig. 19.7 The K^+/π^+ and K^-/π^- ratios as functions of p_T for central (0–5%) and peripheral (60–92%) Au+Au collisions at $\sqrt{s_{NN}} = 200$ GeV (from Ref. [9]).

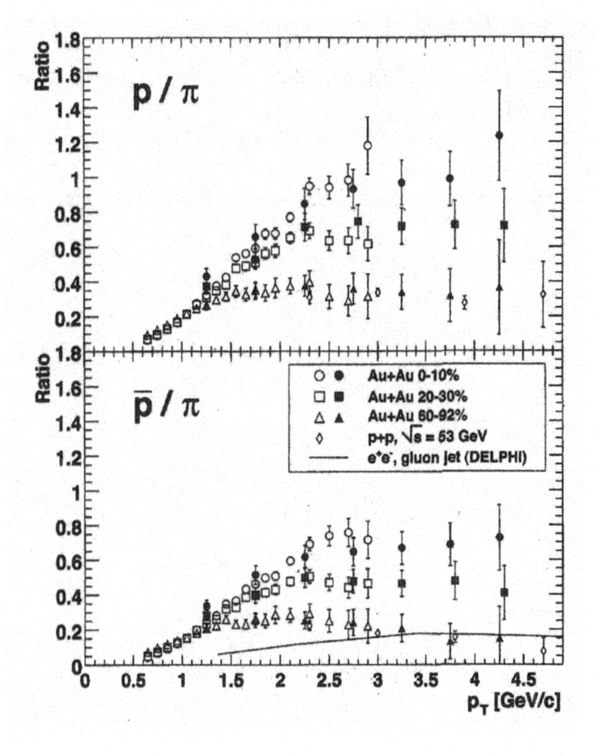

Fig. 19.8 The p/π and $\bar{\mathrm{p}}/\pi$ ratios for central (0–10%), semicentral (20–30%), and peripheral (60–92%) Au+Au collisions at $\sqrt{s_{NN}} = 200$ GeV. Open/filled points are for charged/neutral pions. The data for p+p collisions at $\sqrt{s} = 53$ GeV are also shown. The Figure is from Ref. [9].

References

[1] C. Alt *et al.* (NA49 Collaboration), *Nucl. Phys. A* **774** (2006) 473.
[2] A. Dainese *et al.* (NA57 Collaboration), *Nucl. Phys. A* **774** (2006) 51.
[3] K. Reygers *et al.* (WA98 Collaboration), *J. Phys. G: Nucl. Part. Phys.* **34** (2007) S797.
[4] S. S. Adler *et al.* (PHENIX Collaboration), *Phys. Rev. Lett.* **91** (2003) 072303.
[5] S. S. Adler *et al.* (PHENIX Collaboration), *Phys. Rev. C* **69** (2004) 034910.
[6] B. B. Back *et al.* (PHOBOS Collaboration), *Phys. Rev. Lett.* **91** (2003) 072302.
[7] B. B. Back *et al.* (PHOBOS Collaboration), *Phys. Lett. B* **578** (2004) 297.
[8] J. Adams *et al.* (STAR Collaboration), *Phys. Rev. Lett.* **91** (2003) 072304.
[9] S. S. Adler *et al.* (PHENIX Collaboration), *Phys. Rev. C* **69** (2004) 34909.
[10] K. J. Eskola *et al.*, *Nucl. Phys. A* **747** (2005) 511.
[11] N. Armesto *et al.*, *Phys. Lett. B* **637** (2006) 362.
[12] J. Lahoie *et al.* (PHENIX Collaboration), *J. Phys. G: Nucl. Part. Phys.* **34** (2007) S191.
[13] T. M. Larsen *et al.* (BRAHMS Collaboration), *J. Phys. G: Nucl. Part. Phys.* **34** (2007) S603.
[14] I. Arsene *et al.* (BRAHMS Collaboration), *Phys. Rev. Lett.* **93** (2004) 242303.
[15] J. Casalderrey-Solana and C. A. Salgado, *Acta Phys. Polon. B* **38** (2007) 3731.

Chapter 20

Production and Absorption of Jets

Study of jets in high energy collisions is of great interest as within the frame of QCD jets are believed to result from quarks or gluons, and thus should carry information about the deconfined state of matter, supposedly produced in such reactions.

A jet is defined as a group of particles emitted with close vector momenta, or, in another language, having small relative distances in momentum space. In what follows we will consider hadronic jets, in which "particles" mean final state hadrons. A standard procedure to find a jet is to select a high p_T hadron as a "leading particle", and then look for other hadrons emitted close to it in momentum space. A certain degree of collimation must be assumed in this search, restricting the acceptable jet size. This is of particular importance in high multiplicity events, in order to avoid overlap of different jets. Various jet-finding algorithms differ somewhat between themselves, but they lead to a similar final result. A jet is conveniently visualized using a "lego" plot in the rapidity y (or pseudorapidity η), and azimuthal angle ϕ coordinates. In these coordinates the position of the leading hadron is characterized by some values (η_0, ϕ_0), and the position of the jet axis by somewhat different values (η_c, ϕ_c), obtained as E_T-weighted average taken for all particles belonging to the jet. In such "lego" plot a jet appears as a cone with the base of radius R, with all particles i belonging to the jet fulfilling the relation

$$\sqrt{(\eta_c - \eta_i)^2 + (\phi_c - \phi_i)^2} \leq R \tag{20.1}$$

It should be noted that jet finding is an iterative procedure, much time consuming in multiparticle environment. One requires this procedure to yield relatively stable values of jet parameters (position and radius). Once a jet has been identified, its kinematical characteristics, such as total momentum and effective mass, can be easily calculated from momenta of the selected hadrons.

Formation of jets is typically a high energy process, barely visible at SPS energies, but accessible for investigations at RHIC. Jet studies in nuclear collisions at RHIC led to a very interesting observation. It has been found that jets in the opposite hemisphere (*away-side jets*) show a very different pattern in d+Au and in central Au+Au collisions. This is shown in Fig. 20.1. In d+Au collisions, similarly as in p+p collisions, a pronounced

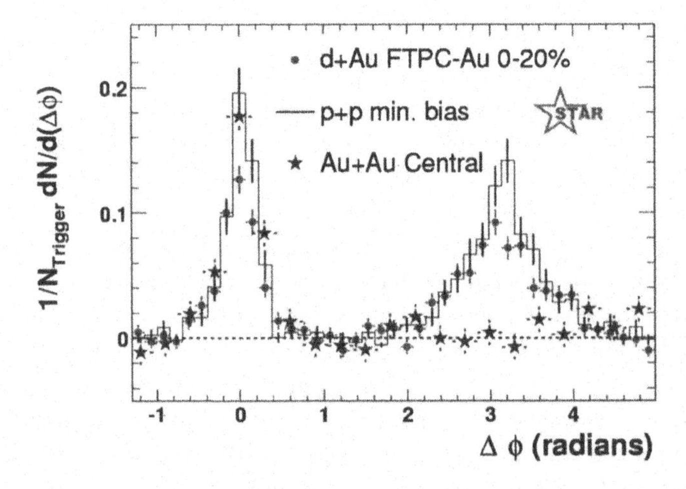

Fig. 20.1 Away-side jet modification in central Au+Au collisions at $\sqrt{s_{NN}} = 200$ GeV with respect to p+p and d+Au collisions at the same energy (from Ref. [1]).

away-side jet appears around $\Delta\phi = \pi$, exactly opposite to the trigger jet, what is typical for di-jet events. In central Au+Au collisions the away-side jet is suppressed. Figure 20.2 shows the evolution of this pattern with centrality of Au+Au collisions. With increasing centrality, the central peak at the away side disappears, and new enhancements of particle yields develop. Figure 20.3 shows this in the azimuthal angle plane: the two side maxima occur symmetrically at the azimuthal angle $\phi \approx 110^o$ with respect to the axis of the trigger jet. Further investigations, mainly studies of three-particle correlations, show the "conical" emission of hadrons in the away-side hemisphere. This is clearly seen in Fig. 20.4 which shows a three-dimensional view of particle emission. A ring-shaped structure appears at the away side. In this Figure the conical profile of the trigger jet, which should be hidden below the plane, has been reflected upwards, and appears

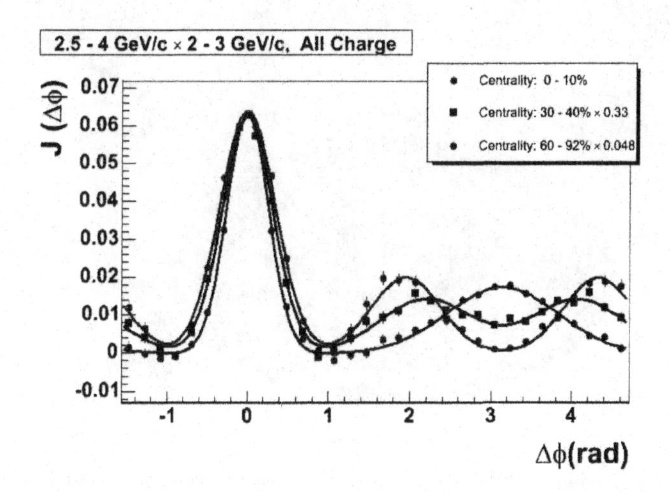

Fig. 20.2 Azimuthal correlation of the trigger jet with associated particles at the away-side in central, mid-central, and peripheral Au+Au collisions at $\sqrt{s_{NN}} = 200$ GeV (from Ref. [2]).

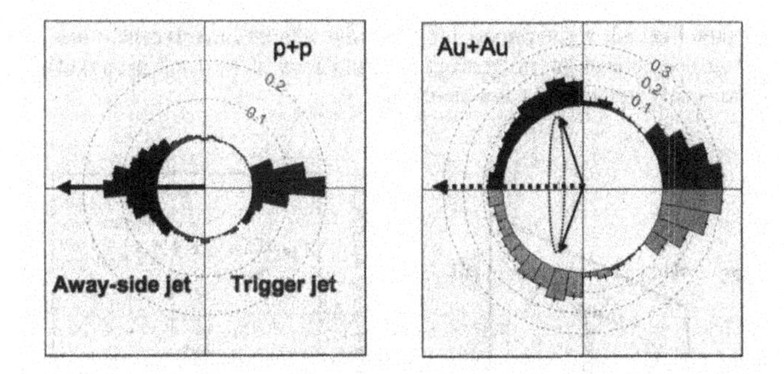

Fig. 20.3 Angular distribution of secondary particles around the beam axis with respect to the direction of the trigger jet in p+p (left panel), and central Au+Au (right panel) collisions at $\sqrt{s_{NN}} = 200$ GeV (from Ref. [3]).

in the centre of the ring structure. The "conical" emission of particles is suggestive of a *Mach cone*.[a]

[a]The Mach cone is formed when an object is moving in a medium (gas or liquid) with velocity which exceeds the velocity of sound in this medium. It separates the undisturbed part of the medium from the disturbed one. The opening angle α of the Mach cone, called the *Mach angle*, is determined by the ratio of the velocity of sound in the medium c to the velocity v of the moving object: $\sin\alpha = c/v$.

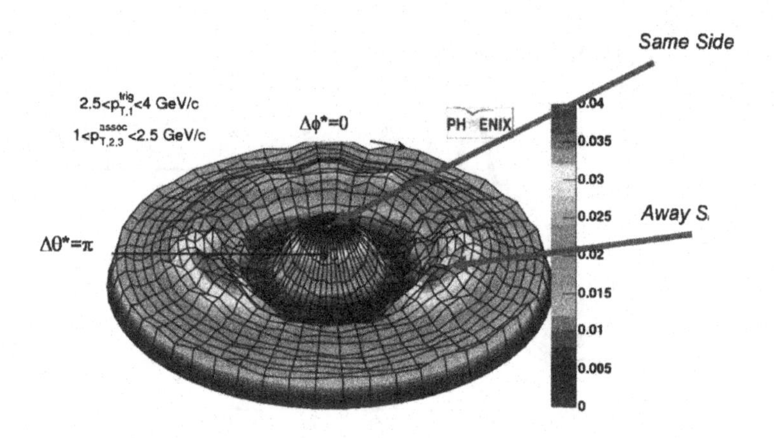

Fig. 20.4 Three-particle correlation function in central Au+Au collisions at $\sqrt{s_{NN}} =$ 200 GeV. The central cone has been reflected upwards (from Ref. [4]).

Formation of the Mach cone indicates that sound propagation in the medium is only weakly damped, what means that the medium viscosity is small.

The observed absorption of jets in the dense medium, also called *jet quenching*, and its dependence on the centrality, or, equivalently, on the geometry of the collision, suggests a possibility of *jet tomography* as a tool to investigate properties of the matter created in the reaction.

References

[1] J. Adams *et al.* (STAR Collaboration), *Nucl. Phys. A* **757** (2005) 102.
[2] Y. Akiba (PHENIX Collaboration), *Nucl. Phys. A* **774** (2006) 403.
[3] B. Müller, *Acta Phys. Polon. B* **38** (2007) 3705.
[4] J. Lahoie (PHENIX Collaboration), *J. Phys. G: Nucl. Part. Phys.* **34** (2007) S191.

Chapter 21

More About Quark-Gluon Plasma

21.1 Polarization of the quark-gluon plasma in the spin space

In non-central collisions of high energy heavy ions large orbital angular momenta are involved. The angular momentum vector is directed perpendicularly to the reaction plane, this being defined by the beam direction and the impact parameter. A fraction of this angular momentum is carried away by non-interacting nuclear fragments — the spectators, but the matter in the overlapped region will also carry a substantial orbital angular momentum. Estimates show that in a non-central Au+Au collision at RHIC energies the global angular momentum of the overlapping matter might be of the order of 10^5 spin units [1]. If quark-gluon plasma is produced, one can speak of a *spinning plasma.* This should result in a global polarization of quarks and antiquarks, and further translate into polarization of the produced spin non-zero hadrons. In particular, polarization of vector mesons and hyperons can be expected. This would be the cleanest signal of a quark-gluon plasma with large intrinsic angular momentum. If hyperons are produced via quark recombination, $q\,q\,q \to Y$, and polarization of all types of quarks is the same, $P_u = P_d = P_s = P_q$, then the polarization of a hyperon would be equal to the polarization of a quark, $P_Y = P_q$. On the other hand, if hyperons are produced via quark fragmentation, $q \to Y + X$, their polarization would be smaller: under the assumption $P_s = P_q$ one obtains $P_Y = P_q/3$. The spin state of a vector meson is described by the spin density matrix, and its diagonal elements measure the amount of spin alignment (without its direction). The vector meson spin alignment is, however, for both quark recombination and quark fragmentation hadronization scenarios, a P_q^2 effect, and thus may be small.

Existence of the global angular momentum of the system also implies a final transverse gradient of the longitudinal flow velocity. An enhancement of the radial and elliptic flow can be expected. Thus, investigating spin effects in non-central collisions of relativistic heavy ions one can obtain interesting information on the reaction evolution, and possibly also on hadronization scenarios.

An attempt to measure spin effects in Au + Au collisions at $\sqrt{s_{NN}} =$ 200 GeV has been made by the STAR Collaboration. For two vector meson species: the $K^{*0}(892)$, and the $\phi(1020)$, the diagonal spin density matrix elements have been determined, and found compatible with no polarization [2]. For hyperons only an upper limit for $\Lambda(\bar{\Lambda})$ global polarization $|P_{\Lambda,\bar{\Lambda}}| <$ 0.02 has been obtained [3]. One should, however, keep in mind that not all Λ hyperons are directly produced, and the feed-down processes would dilute the polarization effects.

In conclusion, no global spin effects have been found in experiment.

21.2 Disoriented chiral condensate

Orientation of the isospin vector of a particle, $\vec{I}$, or of a system of particles, in the isospin space determines its electrical charge which is related to the third component of this vector, I_z. Due to isospin conservation, pions emitted from an isospin zero source should be produced in the three charge states with equal probabilities: $\langle N_{\pi^0}\rangle = \langle N_{\pi^+}\rangle = \langle N_{\pi^-}\rangle$. An isospin $\frac{1}{2}$ pion emitter has a 2/3 probability of emitting a charged pion, and a 1/3 probability of emitting a neutral pion. The mean number of neutral pions is thus $\langle N_{\pi^0}\rangle = \frac{1}{3}\langle N_\pi\rangle$, where $N_\pi = N_{\pi^+} + N_{\pi^-} + N_{\pi^0}$, and their multiplicity distribution is described by the binomial distribution. For large multiplicities this distribution is relatively narrow, as shown in Fig. 21.1 for $N_\pi = 75$. For a multi-hadron system produced in a collision of relativistic heavy ions the situation may, however, be different.

In the pioneering work on this subject [4] it was remarked that when considering the production of classical pion fields, groups of pions can be emitted with close momenta, and within such a group the number of emitted π^0 mesons will in general be different from that of π^+ (or π^-). It was shown that such a state appears in both linear and non-linear σ-models which are simplified versions of the full chiral effective theory [5]. It may have a large isospin vector oriented in any direction in isospace, and thus it may be a source of secondary pions with any isospin configuration. Rajagopal and

Wilczek [6] introduced the notion of *disoriented chiral condensate* (DCC) — a metastable state resulting from cooling down the high temperature chiral symmetric phase of a quark-gluon plasma. A DCC state may occupy the full available phase space or only a part of it, and thus it may constitute a source of all secondary pions or only a small fraction of them. Some theoretical models [7, 8] predict *DCC domains* of sizes 3–4 fm in radius, emitting 50–200 pions. Such a source may be situated in any kinematic region of the expanding system, and the pion emission pattern might be statistical or coherent. If the pion emission from DCC is indeed coherent, the pions will be collimated within a limited region of phase space, and will have small relative transverse momenta. In this case one would expect to find "jet-like" structures ("the pion laser" [9]).

The formation of the DCC would result in a large imbalance in the production of charged to neutral pions. Thus one should see events which mainly consist of the charged pions: $N_{\pi^+} \cong N_{\pi^-} \gg N_{\pi^0}$, or of the neutral pions: $N_{\pi^0} \gg N_{\pi^+} \cong N_{\pi^-}$. This seems to come close to the yet unexplained observations of *centauro* and *anti-centauro* events recorded in emulsion chambers exposed to cosmic rays at mountain altitudes [10].[a]

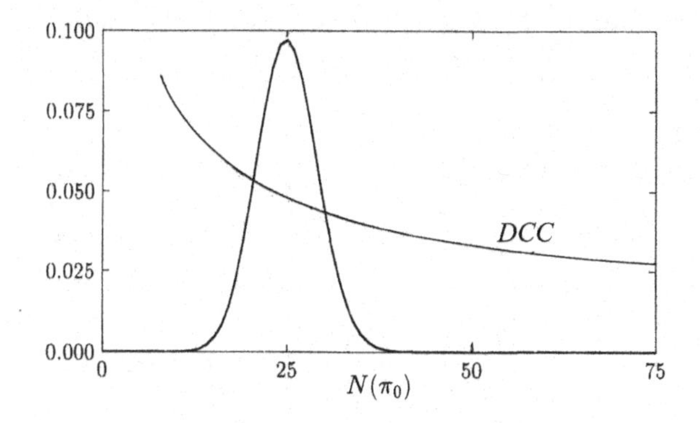

Fig. 21.1 Multiplicity distribution of neutral pions from an isospin zero source, calculated for $N_\pi = 75$ (from Ref. [9]), and analogous distribution expected from DCC (arbitrarily normalized).

The effect of such imbalance between charged and neutral pions can be studied in terms of the distribution of the neutral pion fraction, $f = N_{\pi^0}/N_\pi$. As already stated, neutral pions in a normal event would follow

[a]The *centauro* events are, however, characterized by transverse momenta of the order of 1–2 GeV/c, much higher than those expected from DCC.

a binomial form with a mean of $\frac{1}{3}$, whereas within a domain of DCC the distribution would be $P(f) = 1/(2\sqrt{f})$ [5]. This distribution is also shown in Fig. 21.1. It is very wide, and thus it allows for large fluctuations in the neutral pion fraction. The simplest prescription for finding DCC is thus to count photons and charged hadrons within a common acceptance, and to study the correlation between their numbers. This can be done either in full detector acceptance, or in smaller kinematical regions. Apart of large event-to-event fluctuations in the neutral pion fraction which is the most basic signature of DCC formation, some other signatures have been proposed, and suitable analysis methods developed.

Relevant experimental investigations were performed for Pb+Pb collisions at $\sqrt{s}$ = 17.3 GeV at the CERN SPS. The NA49 Collaboration studied fluctuations in the ratio of electromagnetic to hadronic transverse energy, $E_T^{\rm EM}/E_T^{\rm HAD}$ [11]. The WA98 Collaboration investigated photon and charged particles multiplicity correlations ($N_\gamma - N_{\rm ch}$ correlations) [12], and also made a study using the *discrete wavelet technique* (DWT), which allows to identify fluctuations at various distance scales [13]. No "anomalous" behaviour which might be attributed to a DCC signal has been found.

A search for DCC was also performed for p̄+p collisions at $\sqrt{s}$ = 1.8 TeV at the Tevatron at Fermilab by the dedicated experiment MiniMax [14]. No evidence for DCC was found at a few percent level.

21.3 Color glass condensate

The concept of the *color glass condensate* as a novel state of hadronic matter was motivated by recent results on the gluon density distribution in the nucleon. At the DESY laboratory in Hamburg, Germany, the deep inelastic lepton scattering (DIS) off a hadronic target was investigated. These processes are studied in terms of the *Bjorken variable* $x \approx Q^2/s$, where Q is the four-momentum transfer (or the four-momentum of the exchanged photon), and s is, as usual, the invariant energy squared. Experimental results indicate that at a fixed four-momentum transfer the density of constituents (gluons) in a hadron increases with increasing energy or, equivalently, with decreasing x. The number of gluons is rising rapidly with increasing energy, but as the total cross section rises slowly, the density of gluons should become limited or, in other words, should reach a *saturation.*[b] Saturation occurs for transverse momenta of produced particles below a certain critical

[b]In fact, it would show a slow, logarithmic increase with $1/x$.

value, Q_s^2, called the *saturation scale.* This quantity increases with increasing energy, and can become larger than the QCD scale $\Lambda_{\rm QCD}$, what means that the strong coupling constant α_s decreases, and may become smaller than one, $\alpha_s \ll 1$. The system is thus weakly coupled. This is a new and important feature which would allow to use a perturbative approach to the description of such a system.

This novel state of matter was named *Color Glass Condensate* (CGC) for the reasons given below [15–17]:

- "Color" since the gluons are colored;
- "Glass" because of the strong analogy to the actual glass. A glass is a system with "frozen disorder" which is like a solid at short time scales, and like a liquid at much longer time scales. Similarly, gluons are disordered, and evolve slowly in longitudinal momentum in a manner analogous to a glass;
- "Condensate" because it contains a very high density of gluons. Increasing the energy leads to the saturation of the gluon density, what corresponds to a multiparticle Bose condensate state.

The concept of CGC has been applied to the phenomenology of DIS, and also to that of relativistic heavy ion collisions. One should, however, keep in mind that a smallness of the strong coupling constant α_s requires a rather hard saturation scale: $Q_s^2 \gg \Lambda_{\rm QCD}^2$, and for the presently available energies this condition is only marginally satisfied. In nuclei the saturation scale Q_s increases with A as $Q_s^2 \propto A^{1/3} \ln A$, what makes the conditions for the application of the CGC approach more favourable, but still an estimate of the saturation scale at RHIC gives $Q_s \approx$ 1–2 GeV, and at LHC $Q_s \approx$ 2–3 GeV [16]. These rather moderate values of Q_s mean that the strong coupling constant α_s is in fact not very small, and thus perturbative calculations should not be limited to the lowest order in α_s.

Nevertheless, calculations based on the concept of CGC have yielded some encouraging results. In particular, the feature of *limiting fragmentation* observed in collisions of relativistic nuclei has found a plausible explanation in the framework of CGC [17]. There is hope that more quantitative successes of this approach will be reported.

A new concept, connected to CGC, is that of *glasma.* This is supposed to be a state of matter emerging just after the collision of two ultrarelativistic nuclei, each of them described as a sheet of color glass. Glasma melts into gluons, which thermalize together with quarks, leading to the formation of the quark-gluon plasma [17].

References

[1] Z. Liang, *J. Phys. G: Nucl. Part. Phys.* **34** (2007) S323.
[2] J. H. Chen *et al.* (STAR Collaboration), *J. Phys. G: Nucl. Part. Phys.* **35** (2008) 044068.
[3] B. Abelev *et al.* (STAR Collaboration), *Phys. Rev. C* **76** (2007) 024915.
[4] A. A. Anselm and M. G. Ryskin, *Phys. Lett. B* **266** (1991) 482.
[5] J. P. Blaizot and A. Krzywicki, *Phys. Rev. D* **46** (1992) 246.
[6] K. Rajagopal and F. Wilczek, *Nucl. Phys. B* **399** (1993) 395.
[7] S. Gavin *et al.*, *Phys. Rev. Lett.* **72** (1994) 2143; S. Gavin and B. Müller, *Phys. Lett. B* **329** (1994) 486.
[8] J. I. Kapusta and A. P. Vischer, *Z. Phys. C* **75** (1997) 507.
[9] S. Pratt, *Phys. Lett. B* **301** (1993) 159.
[10] C. M. G. Lates, Y. Fujimoto and S. Hasegawa, *Phys. Rep.* **65** (1980) 151.
[11] P. Seyboth, *Proc. XXV Int. Symposium on Multiparticle Dynamics*, Stara Lesna, Slovakia, 1995 (World Scientific, 1996), p. 170.
[12] M. M. Aggarwal *et al.* (WA98 Collaboration), *Phys. Lett. B* **420** (1998) 169; *Phys. Rev. C* **67** (2003) 044901.
[13] M. M. Aggarwal *et al.* (WA98 Collaboration), *Phys. Rev. C* **64** (2001) 011901; *ibid.* **67** (2003) 044901.
[14] T. C. Brooks *et al.*, *Phys. Rev. D* **55** (1997) 5667.
[15] E. Iancu, *Nucl. Phys. A* **715** (2003) 219c.
[16] E. Iancu and R. Venugopalan, *Quark-Gluon Plasma 3*, eds. R. Hwa and X.-N. Wang (World Scientific, 2004), p. 249.
[17] L. McLerran, *Nucl. Phys. A* **787** (2007) 1c.

Chapter 22

Predictions for the Large Hadron Collider

22.1 Extrapolations of present-day experimental data

Near the end of this year the first proton–proton collisions are expected in the Large Hadron Collider (LHC) now in the final stage of commissioning at CERN, and nuclear beams are expected in the following year. For Pb+Pb collisions the centre-of-mass (c.m.) energy per nucleon pair will be $\sqrt{s_{NN}} = 5.5$ TeV, almost 30 times higher than that in RHIC. For p+p collisions the c.m. energy will reach $\sqrt{s} = 14$ TeV. This will be a large leap in the collision energy, nevertheless some predictions related to various characteristics of elementary and nuclear collisions at the LHC can be made using simple extrapolation of the present-day experimental data. Data suitable for such extrapolations come mainly from experiments at RHIC, and in the first place from the PHOBOS experiment [1].

As already mentioned in Chapter 8, the total multiplicity of charged secondary particles in relativistic nuclear collisions increases with the energy of the collision as $\ln^2 \sqrt{s_{NN}}$, while the density of charged secondaries at midrapidity increases as $\ln \sqrt{s_{NN}}$. In Figs. 22.1 and 22.2 these relations are used for extrapolation to LHC energies.

Figure 22.1 presents the logarithmic extrapolation of the total charged multiplicity per participant pair. Two classes of p+p collisions: inelastic and non-single-diffractive (NSD), together with nuclear collisions, are shown. The obtained numerical values at LHC energies are collected in Table 22.1.

Figure 22.2 shows the energy dependence of the mean density of charged secondary particles at midrapidity for nuclear collisions, again per participant pair. The logarithmic extrapolation yields the value

$$dN_{\text{ch}}/d\eta\,|_{\eta=0} = 1200 \pm 100$$

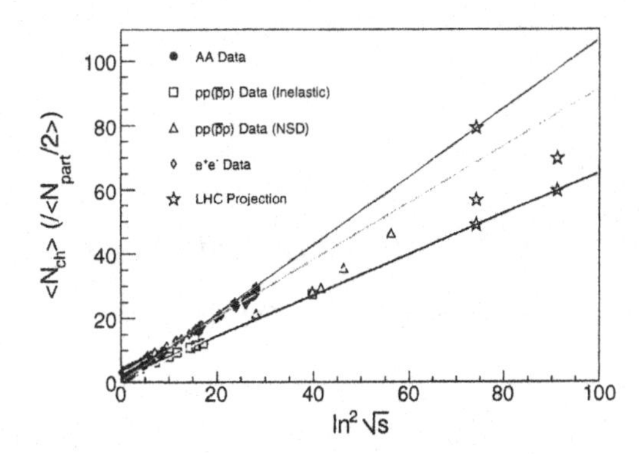

Fig. 22.1 Energy dependence of secondary charged particles multiplicities per participant pair in various reactions, extrapolated to LHC energies. Straight lines indicate the $\propto \ln^2 \sqrt{s_{NN}}$ behaviour (from Ref. [2]).

Table 22.1 Mean multiplicities of charged secondary particles for p+p and for central Pb+Pb collisions extrapolated to LHC energies (from Ref. [2]).

Reaction	$\sqrt{s_{NN}}$, TeV	$\langle N_{ch} \rangle$
p+p inel.	5.5	49 ± 8
p+p NSD	5.5	57 ± 7
p+p inel.	14.0	60 ± 10
p+p NSD	14.0	70 ± 8
Pb+Pb central	5.5	15000 ± 1000

for the most central (top 3% centrality) Pb+Pb collisions at $\sqrt{s_{NN}} = 5.5$ TeV. Charged particle multiplicities at SPS and RHIC energies can also be fitted reasonably well with a power law in $\sqrt{s}$ with the exponent of about 0.29 [3]. Using it for extrapolation to LHC energy one arrives at the values which are higher than for the logarithmic extrapolation: $dN_{ch}/d\eta|_{\eta=0} =$ 1670–1900 for most central ($N_{part} = 375$) Pb+Pb collisions at $\sqrt{s_{NN}} =$ 5.5 TeV.

Figure 22.3 shows the density distribution of charged secondaries, $dN_{ch}/d\eta$, in central (0–6% centrality) and mid-central (35–45% centrality) Pb+Pb collisions at $\sqrt{s_{NN}} = 5.5$ TeV, extrapolated from the PHOBOS Au+Au results at RHIC. One should, however, mention another possible extrapolation of particle density distributions to LHC energies: extending the "limiting fragmentation" regions up to midrapidity. This would result

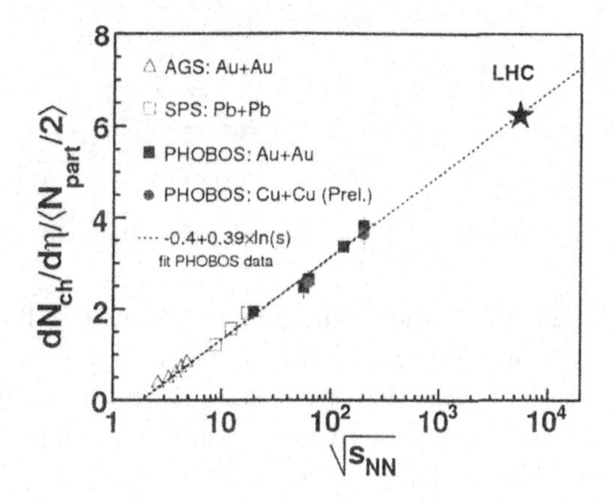

Fig. 22.2 Energy dependence of the midrapidity density of charged secondary particles per participant pair in collisions of various nuclei, extrapolated to the LHC. Straight line indicates the $\propto \ln\sqrt{s_{NN}}$ behaviour (from Ref. [2]).

in a triangular rather than trapezoidal shape of the distribution, and lead to a higher value of particle density at midrapidity: $dN_{\text{ch}}/d\eta\,|_{\eta=0} \cong 1700$ [4], instead of the above quoted value of 1200.

Figure 22.4 shows the expected density distribution of "net protons" (i.e. "protons minus antiprotons"), calculated on the assumption that the nucleon rapidity loss at LHC is $\Delta y \approx 2$, almost the same as at RHIC energies [5] (see Fig. 9.4 of Chapter 9).

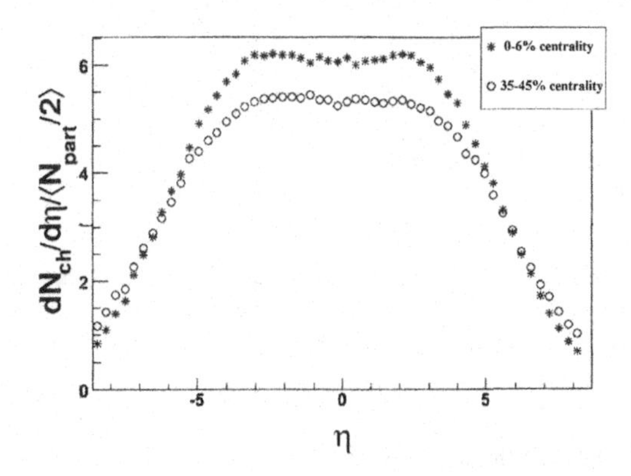

Fig. 22.3 Pseudorapidity density distribution of charged secondary particles per participant pair for Pb+Pb collisions with two different centralities, extrapolated from RHIC to LHC energies (from Ref. [2]).

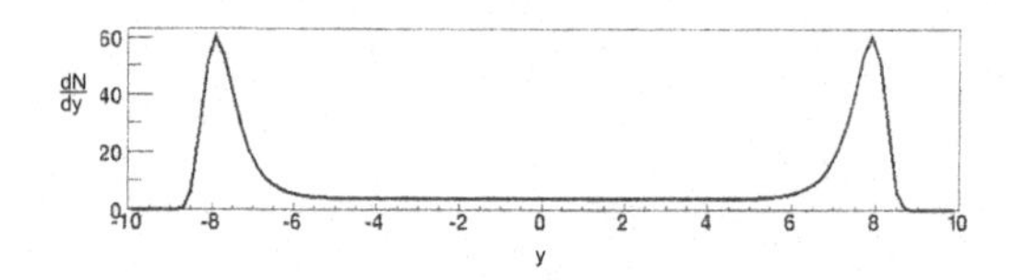

Fig. 22.4 Rapidity distribution of "net protons" expected at LHC (from Ref. [6]).

Finally, Fig. 22.5 shows the predictions for the elliptic flow. The pseudorapidity distribution of the parameter v_2 which characterizes the elliptic flow (see Chapter 15) for the 40% most central Pb+Pb collisions at LHC has been extrapolated from the PHOBOS results, assuming that the triangular shape of the distribution observed at RHIC (see Fig. 15.5 of Chapter 15) will persist at LHC, and that the distribution will show similar slope at both edges (the "limiting fragmentation" feature). These assumptions define the distribution, and lead to a prediction that v_2 at midrapidity at LHC might be larger than that at RHIC by as much as 50%. For the class of events considered here as example (0–40% centrality) $v_2 \,|_{\eta=0} \cong 0.050$ at RHIC, and $v_2 \,|_{\eta=0} \cong 0.075$ can be expected at LHC.

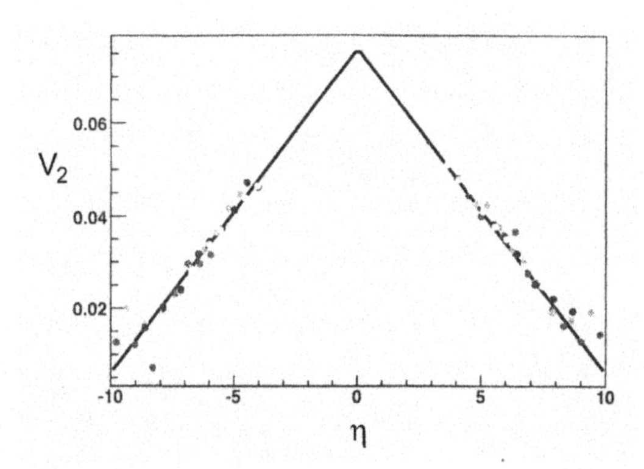

Fig. 22.5 Pseudorapidity distribution of elliptic flow for the 40% most central collisions extrapolated from Au+Au at RHIC to Pb+Pb at LHC energies. Points show the RHIC data at several energies which determine the slope of the distribution at its both edges (from Ref. [2]).

When making predictions for the LHC one should, however, keep in mind that simple scaling properties of various characteristics observed in present-day experiments might break down at higher energies. The main reason is the unknown contribution of "hard" collisions. Studies of p+A collisions indicated that the yield of secondary particles is proportional to the number of nucleons-participants, $N_{\rm part}$.[a] Such situation is typical for "soft" processes,[b] and it also seems to hold for A+A collisions, however not exactly. An attempt to estimate the contribution of "hard" processes in Au+Au collisions at RHIC energies was made by Kharzeev and Nardi [8]. They assumed that the fraction x of the multiplicity per unit of (pseudo)rapidity measured in p+p collisions is due to "hard" processes, and the remaining fraction $(1 - x)$ comes from "soft" processes. Then the multiplicity in nuclear collisions will have two components: "soft", which is assumed to be proportional to the number of participants, $N_{\rm part}$, and "hard" which is proportional to the number of collisions, $N_{\rm coll}$. This is expressed in the formula

$$\frac{dN_{\rm ch}}{d\eta} = N_{\rm ch}^{\rm pp}\left[(1-x)\frac{N_{\rm part}}{2} + x\,N_{\rm coll}\right] \tag{22.1}$$

where $N_{\rm ch}^{\rm pp}$ is the yield of charged secondary particles in p+p collisions at the same collision energy, and x determines the relative contribution of "hard" collisions. At SPS energy x is negligible (only "soft" processes), while at $\sqrt{s_{NN}} = 130$ GeV at RHIC the estimate of Ref. [8] is $x \cong 0.09 \pm 0.03$. Let us note that this would mean a very substantial contribution of "hard" processes to the measured multiplicity: about 37% of particles produced in Au+Au collisions at this energy would result from "hard" processes. The uncertainty of this number is, however, quite large, and the rate of further increase of x with energy is unknown. As "hard" collisions may show different characteristics, and different (pseudo)rapidity dependence as compared to "soft" ones, their increasing contribution may change the overall picture, and invalidate simple extrapolations quoted above.

22.2 Predictions from theoretical models

In order to collect predictions from the theorists, a special workshop entitled *Heavy Ion Collisions at the LHC — Last Call for Predictions* was organized at CERN in May/June 2007 [9, 10]. Predictions presented at this Workshop are issued from various theoretical models and computer codes based on them. Such models are, in general, not parameter-free, and

[a] These are nucleons which suffered the first collisions, or "wounded" nucleons in the terminology of Białas *et al.* [7], see also Chapter 6.

[b] Yield from "hard" processes is expected to be proportional to the number of binary collisions, $N_{\rm coll}$.

have some parameters which are adjusted so as to fit the present-day experimental data. With some energy dependence of these parameters, the models also have elements of an extrapolation from lower to higher energies. We shall quote some selected predictions.

The mean multiplicity of charged secondary particles calls a special attention. This is mainly because it is the "first day observable" which can discriminate between different theoretical models. Also, many other predictions depend on it. The early compilation of theoretical predictions on the charged particle density at midrapidity, $dN_{\mathrm{ch}}/d\eta|_{\eta=0}$, for the LHC dated from before the commissioning of RHIC, quoted values ranging from about 1200 up to 8000, a very wide interval [11]. More recent predictions are situated in the lower part of this interval, as examples we quote: 2000 from a CGC approach [12], 2570 from perturbative QCD + hydrodynamics [13], and 3500 from the HIJING code [14].

The multiplicity distribution in rapidity (or pseudorapidity) is usually assumed to be flat, or almost flat, in the central region ("plateau"), however the DPMJET code, based on the Dual Parton Model [15], predicts for Pb+Pb central collisions at $\sqrt{s_{NN}} = 5.5$ TeV a pronounced bimodal ("double-humped") structure with $dN_{\mathrm{ch}}/d\eta \approx 1800$ at $\eta = 0$, and $dN_{\mathrm{ch}}/d\eta \approx 2000$ at $\eta = \pm 2$.

Spectra of identified particles at midrapidity in Pb+Pb collisions at $\sqrt{s_{NN}} = 5.5$ TeV have been studied using hydrodynamic [16] and combined hydrodynamic and perturbative QCD (pQCD) [17] approaches. Generally, all spectra at the LHC are expected to be less steep than at RHIC due to an increased radial flow. For low transverse momenta the spectra are described by relativistic hydrodynamics, and for higher p_T they are predicted to become flatter due to contributions from (mini)jet fragmentation. For central collisions the hydrodynamic contribution dominates over the pQCD one up to $p_T \approx 4$ GeV/c, and for peripheral collision only up to $p_T \approx 1.5$ GeV/c. The hydro-pQCD p_T crossing point moves towards higher p_T with increasing hadron mass. The rise of the ratios of heavy-to-light hadrons (e.g. $\bar{\mathrm{p}}/\pi$, Λ/K) with p_T will be slower at LHC than at RHIC.

Ratios of various hadronic species can be predicted from the statistical-thermal model which turned out to be very successful at present-day energies [18]. Let us recall that some particle ratios exhibit a large sensitivity to the thermal parameters, in particular the antiparticle-to-particle ratios strongly depend on the baryonic chemical potential. At the LHC the expected values of the temperature and baryonic chemical potential are $T = 161 \pm 4$ MeV, and $\mu_B = 0.8^{+1.2}_{-0.6}$ MeV. The antiparticle/particle ratios will be all very close to unity, except for the $\bar{p}/p$ ratio which is predicted to be 0.95, reflecting the expected small, but non-zero, value of μ_B.

Two-pion and two-kaon correlations have been calculated using the computer code which takes into account their final-state strong and Coulomb interactions [19], and which reproduces the RHIC data reasonably well. Table 22.2 shows the Bertsch radii obtained from Gaussian fits to correlation functions for low p_T pions and kaons emitted near midrapidity in central Au+Au collisions at $\sqrt{s_{NN}}$ = 200 GeV (RHIC), and similar radii predicted for central Pb+Pb collisions at $\sqrt{s_{NN}}$ = 5.5 TeV (LHC). One can see that the radii at the LHC are larger than those at RHIC (the correlation functions will be narrower). Radii of the emission source for kaons are smaller than those for pions. The $R_{\text{out}}/R_{\text{side}}$ ratio is in all cases close to one. Similar results on the pion source radii have been obtained with

Table 22.2 Radii from Gaussian fits to correlation functions at RHIC and predictions for the LHC (from Ref. [9]).

	R_{out}, fm	R_{side}, fm	R_{long}, fm	λ
pions, RHIC	3.60	3.52	3.23	0.50
pions, LHC	4.23	4.70	4.86	0.43
kaons, RHIC	2.95	2.79	2.62	0.94
kaons, LHC	3.56	3.20	3.16	0.89

the *Fast Hadron Freeze-out Generator* of Ref. [20]. Calculations using the relativistic hydrodynamics result in much weaker energy dependence of the correlation radii [21].

Relativistic hydrodynamics allows to obtain predictions concerning the radial and elliptic flow. The differential elliptic flow, $v_2(p_T)$, for $p_T < 1.5$ GeV/c should decrease with increasing energy, but both radial and p_T-integrated elliptic flow are predicted to increase from RHIC to LHC [16]. This is due to the relative depletion of low p_T hadrons, and a shift of the momentum asymmetry to higher values of p_T. The p_T-integrated elliptic flow for pions is expected to increase from RHIC to LHC by about 25% [22].

The J/ψ suppression is predicted in Ref. [23] to increase by a factor of 5–6 as compared to RHIC.

The suppression of the away-side jet in Au+Au central collisions at RHIC, and the appearance of the Mach cone, is commonly interpreted as jet energy loss, or *jet quenching*. However, the interaction of a jet in the medium is theoretically not well understood, and there are significant uncertainties in the energy loss models. They affect both the explanation of the RHIC data, and the extrapolation to the LHC.

For similar reasons, predictions concerning nuclear modification factors R_{AA} and R_{CP} also have large uncertainties.

Possibilities of observation of some "exotic" phenomena at the LHC are also being discussed. These are: the quark-gluon plasma with high intrinsic angular momentum (in peripheral collisions), the disoriented chiral condensate and *centauro*-like events, production of exotic multi-quark systems like charmed tetra-quarks and penta-quarks, and generation of *mini black holes.* More details on these topics can be found in Ref. [9].

References

[1] B. B. Back *et al.* (PHOBOS Collaboration), *Nucl. Phys. A* **757** (2005) 28.
[2] W. Busza, *J. Phys. G: Nucl. Part. Phys.* **35** (2008) 044040.
[3] N. Armesto, C. A. Salgado and U. A. Wiedemann, *Phys. Rev. Lett.* **94** (2005) 022002.
[4] U. A. Wiedemann, *J. Phys. G: Nucl. Part. Phys.* **34** (2007) S503.
[5] I. G. Bearden (BRAHMS Collaboration), *J. Phys. G: Nucl. Part. Phys.* **34** (2007) S207.
[6] H. H. Dalsgaard (BRAHMS Collaboration), Poster No. 6 at *Quark Matter'06*, Shanghai, 2006.
[7] A. Białas, M. Błeszyński and W. Czyż, *Nucl. Phys. B* **111** (1976) 461.
[8] D. Kharzeev and M. Nardi, *Phys. Lett. B* **507** (2001) 121.
[9] *Proc. CERN Workshop Heavy Ion Collisions at the LHC — Last Call for Predictions*, preprint arXiv:hep-ph/0711.0974v1 (2007).
[10] N. Armesto, *Proc. Quark Matter 2008*, to appear in *J. Phys. G: Nucl. Part. Phys.*
[11] N. Armesto and C. Pajares, *Int. J. Mod. Phys. A* **15** (2000) 2019.
[12] D. Kharzeev, E. Levin and M. Nardi, *Nucl. Phys. A* **747** (2005) 609.
[13] K. J. Eskola *et al.*, *Phys. Rev. C* **72** (2005) 044904.
[14] V. Topor Pop *et al.*, in Ref. [9].
[15] S. Roesler, R. Engel and J. Ranft, *Nucl. Phys. B (Proc. Suppl.)* **122** (2003) 392.
[16] G. Kestin and U. Heinz, arXiv:0806.4539v2 (nucl. th.), 2008.
[17] F. Arleo, D. d'Enterria and D. Peressounko, in Ref. [9].
[18] A. Andronic, P. Braun-Munzinger and J. Stachel, *Nucl. Phys. A* **772** (2006) 167.
[19] S. Pratt *et al.*, *Nucl. Phys. A* **566** (1994) 103c.
[20] N. S. Amelin *et al.*, *Phys. Rev. C* **74** (2006) 064901.
[21] E. Frodermann, R. Chatterjee and U. Heinz, *J. Phys. G: Nucl. Part. Phys.* **34** (2007) 2249.
[22] T. Hirano *et al.*, *J. Phys. G: Nucl. Part. Phys.* **34** (2007) S879.
[23] A. Capella and C. G. Ferreiro, in Ref. [9].

Appendix A

Relativistic Kinematics

A.1 Basic definitions and formulae

A particle is called *relativistic* if its energy is comparable or exceeds its rest energy mc^2 (if $E \gg mc^2$ one sometimes speaks of *ultrarelativistic* particle.)

In relativistic physics the following relation holds between the total energy E and momentum p of a particle with rest mass m:

$$E^2 = (pc)^2 + (mc^2)^2 \tag{A.1}$$

where c is the velocity of light. It is convenient to use the system of units in which c=1. Then the relation Eq. (A.1) becomes simply

$$E^2 = p^2 + m^2 \tag{A.2}$$

where all three quantities are expressed in units of energy, e.g. in GeV (1 GeV=10^9eV). Inspite of this generally adopted convention, the momentum of a particle is usually quoted in GeV/c while energy and mass are both quoted in GeV. Thus a "1 GeV/c particle" is a particle having momentum of 1 GeV/c, while a "1 GeV particle" is a particle having energy of 1 GeV, and a possible misunderstanding is avoided.

Introducing the Lorentz factor $\gamma = (1 - \beta^2)^{-1/2}$ where β is the velocity in units of the velocity of light, $\beta = v/c$, one finds that the following simple relations hold: $E = \gamma m$, $p = \gamma\beta m$, $p = \beta E$, with $\beta \leq 1, \gamma \geq 1$. For an ultrarelativistic particle $\gamma \gg 1$, $\beta \approx 1$, and $E \approx p$.

Relativistic kinematics is based on the Lorentz transformation which relates energy and momentum components of a particle in a given reference frame to those in another frame moving with relative velocity β_f

$$\begin{pmatrix} E^* \\ p_L^* \end{pmatrix} = \begin{pmatrix} \gamma_f & -\gamma_f\beta_f \\ -\gamma_f\beta_f & \gamma_f \end{pmatrix} \begin{pmatrix} E \\ p_L \end{pmatrix} \;, \; p_T^* = p_T \tag{A.3}$$

where p_L, p_T are the components of $\vec{p}$ parallel and perpendicular to β_f, and $\gamma_f = (1 - \beta_f^2)^{-1/2}$. From this one finds for the longitudinal momentum component in the moving frame

$$p_L^* = \gamma_f \left(p_L - \beta_f E\right) \tag{A.4}$$

and for the corresponding energy

$$E^* = \gamma_f \left(E - \beta_f p_L\right) \tag{A.5}$$

A.2 Rapidity and pseudorapidity

A very useful variable is *rapidity* defined as

$$y = \tanh^{-1}\beta = \frac{1}{2}\ln\left(\frac{1+\beta}{1-\beta}\right) \tag{A.6}$$

The relation between y and β is plotted in Figure A.1.

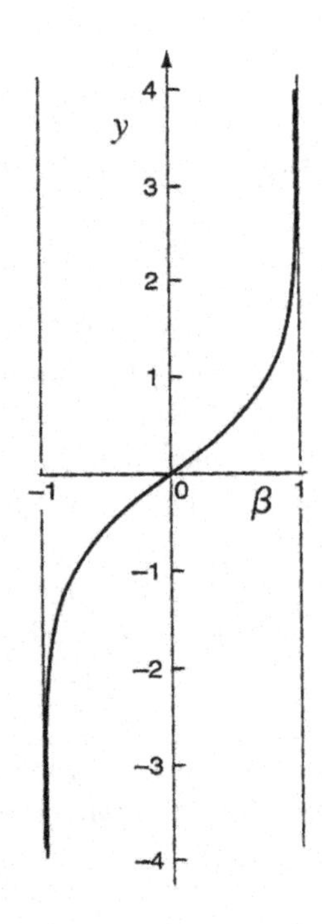

Fig. A.1 Rapidity plotted against velocity.

One can see that $y \approx \beta$ for small β, while $y \to \infty$ for $\beta \to 1$. Equation (A.6) can be rewritten in terms of other variables

$$y = \frac{1}{2}\ln\left(\frac{E+p_L}{E-p_L}\right) \tag{A.7}$$

or

$$y = \ln\left(\frac{E+p_L}{m_T}\right) \tag{A.8}$$

where $m_T = \sqrt{m^2 + p_T^2}$ is called *transverse mass.* Rapidity is additive under Lorentz transformation: $y' = y - \tanh^{-1}\beta_f$. In particular, rapidity of a particle in the centre-of-mass (c.m.) frame, y^*, is related to its rapidity in the laboratory frame, y, by the relation $y^* = y - \tanh^{-1}\beta_c$ where β_c is the velocity of the c.m.frame. This also means that the shape of the rapidity distribution $d\sigma/dy$ is invariant under Lorentz transformation, and acquires only a parallel shift when going from one reference frame to another. For a collision of two equal-mass objects (can be particles or nuclei), the rapidity of the c.m.frame equals one half of that of the laboratory frame, $y_c = 0.5\,y_{\text{lab}}$. For ultrarelativistic particles $E \approx p$, and with $\cos\theta = p_L/p$, where θ is the emission angle, one obtains $E \pm p_L \approx E(1 \pm \cos\theta)$, and rapidity can be approximated by *pseudorapidity*

$$\eta = -\ln\tan(\theta/2) \qquad \text{(A.9)}$$

Using η allows one to obtain the particle distribution in the centre-of- mass frame by measuring only angles in the laboratory frame. This has been used since long in investigations of cosmic ray interactions and has also been proven very useful in experimental set-ups without magnetic field, where only angles are measured. One should, however, remember that replacing rapidity by pseudorapidity provides a faitly good approximation for pions, but is exact only for photons.

A.3 Scaled variables

The Feynman's x-variable, usually denoted as x_F, is defined as the fraction of the maximum allowed longitudinal momentum in the c.m.frame which is carried by a given particle, $x_F = p_L^*/p_{L\max}^*$. Obviously $-1 \leq x_F \leq 1$. This variable is used e.g. for comparing the shapes of particle distributions at different incident energies.

Scaling properties of various distributions can also be studied in terms of the scaled rapidity $z = y^*/y_{\max}^*$. The two scaled variables: x_F and z emphasize different kinematic regions: the detailed structure of the central part of the distribution (i.e. large emission angles) can be better seen in x_F, while the far "wings" (i.e. small angles) in z.

A.4 Invariant mass and centre-of-mass energy

Invariant mass of a group of k particles with energies E_1, E_2,...E_k and momenta $\vec{p_1}$, $\vec{p_2}$,...$\vec{p_k}$ is given by

$$M_k^2 = \Big(\sum_{i=1}^{k} E_i\Big)^2 - \Big(\sum_{i=1}^{k} \vec{p_i}\Big)^2 \qquad \text{(A.10)}$$

This formula constitutes the basis for unstable particle search and "resonance hunting": if a bound state or a resonance with mass M occurs in a system of k particles, then the invariant mass distribution $d\sigma/dM_k$ shows a peak at this mass value.

If two particles with energies E_1, E_2 and momenta $\vec{p_1}$, $\vec{p_2}$ collide, then the total centre-of-mass energy represents in fact the invariant mass of the system

$$E_{CM}^2 = M_2^2 = (E_1 + E_2)^2 - (\vec{p_1} + \vec{p_2})^2 \tag{A.11}$$

or

$$E_{CM}^2 = m_1^2 + m_2^2 + 2E_1E_2(1 - \beta_1\beta_2 \cos\theta) \tag{A.12}$$

where θ is the angle between momentum vectors of these particles. If particle 2 is at rest, then $\beta_2 = 0$, $E_2 = m_2$, and

$$E_{CM}^2 = m_1^2 + m_2^2 + 2E_1m_2 \tag{A.13}$$

or for $E_1 \gg m_1, m_2$ simply

$$E_{CM} = (2E_1m_2)^{1/2} \tag{A.14}$$

For ultrarelativistic particles colliding head-on $\cos\theta = -1$, and

$$E_{CM} = 2\,(E_1E_2)^{1/2} \tag{A.15}$$

or

$$E_{CM} = 2E \tag{A.16}$$

if both particles have equal energies. Comparison of Eqs. (A.16) and (A.14) shows the net advantage of a collider where one has the total incident energy at one's disposal as "useful energy" as compared to a single-beam accelerator working on a stationary target. In the latter case the c.m. energy is proportional to the square root of the incident particle energy, what means that in order to gain e.g. a factor of ten in E_{CM} one should build an accelerator providing beams of energies one hundred times higher.

A.5 Decay processes

If a particle of mass M decays into two particles with masses m_1, m_2, then in the rest frame of particle M each of its decay products acquires the momentum

$$p_1^* = p_2^* = \frac{1}{2M}[(M^2 - (m_1 + m_2)^2)(M^2 - (m_1 - m_2)^2)]^{1/2} \tag{A.17}$$

the two decay products being emitted in opposite directions due to momentum conservation.

If both decay products have the same mass, $m_1 = m_2 = m$, then

$$p^* = \frac{1}{2}(M^2 - 4m^2)^{1/2} \tag{A.18}$$

In the case of a decay into three particles, the available energy is distributed among them, what results in continuous energy spectra starting from zero and extending up to the maximum value allowed by energy-momentum conservation. Due to momentum vector conservation the three decay products will be coplanar in the rest frame of the parent particle.

A.6 Invariant cross sections

The invariant triple differential cross section $Ed^3\sigma/dp^3$ can be written in terms of the longitudinal and transverse momentum components p_L, p_T and azimuthal angle ϕ

$$E\frac{d^3\sigma}{dp^3} = E\frac{d^3\sigma}{d\phi\, dp_L\, p_T\, dp_T} = E\frac{2\, d^3\sigma}{d\phi\, dp_L\, d(p_T^2)} \tag{A.19}$$

or, replacing p_L with rapidity y by using the relation $dy/dp_L = 1/E$

$$E\frac{d^3\sigma}{dp^3} = \frac{2\, d^3\sigma}{d\phi\, dy\, d(p_T^2)} \tag{A.20}$$

Averaging over ϕ yields the double differential cross section

$$\frac{d^2\sigma}{\pi\, dy\, d(p_T^2)} \tag{A.21}$$

Integration over y leads to the single differential distribution $d\sigma/d(p_T^2)$ called the transverse momentum squared distribution, and integration over p_T leads to the single differential distribution $d\sigma/dy$ called the rapidity distribution. The invariant shape of the latter has been already discussed earlier in this chapter. In particular, an isotropic angular distribution (e.g. a decay of a thermal fireball placed at $y = y_0$) has in the rapidity variable the shape $dn/dy \sim \cosh^{-2}(y - y_0)$ which is close in shape to a Gaussian with dispersion $\sigma = 0.88$. Using η instead of y slightly modifies the shape of the distribution, mainly in the central region where particles are relatively slower, as

$$\frac{d\sigma}{d\eta} = \frac{1}{\beta}\frac{d\sigma}{dy} \tag{A.22}$$

A.7 Motion of a particle in external fields

A particle with charge q moving with velocity $\vec{v}$ in an external electromagnetic field $(\vec{E}, \vec{B})$ is subject to the Lorentz force

$$\vec{F} = \frac{d\vec{p}}{dt} = q\vec{E} + q\vec{v} \times \vec{B} \tag{A.23}$$

The first term in this formula is the electric force which accelerates the particle (or slows it down). Energy unit used in subatomic physics is based on this formula: 1 electron-volt (eV) is the energy gained by a singly charged particle moving in the electric field between points with the potential difference of 1V. In high energy physics multiples of this basic unit are used: 1 keV = 10^3 eV, 1 MeV = 10^6 eV, 1 GeV = 10^9 eV, 1 TeV = 10^{12} eV, 1 PeV = 10^{15} eV.

The second term in Eq. (A.23) is the magnetic force which acts along the velocity component perpendicular to the direction of the field and makes the particle follow a helicoidal trajectory in space without changing its energy. In the plane containing the magnetic field vector the projected particle trajectory is not affected and remains to be a straight line - this is the so-called *non-bending plane.* In the plane perpendicular to the magnetic field vector (the *bending plane*) the projected particle trajectory is a circle with the radius $\rho = m\,v_T/qB$ where v_T is the velocity component perpendicular to the magnetic field. A useful quantity is the product $B\rho$ called the *magnetic rigidity* of a particle. It is equal to the momentum of the particle divided by its charge: $B\rho = p/q$, and is usually expressed in Tesla.meters (T.m). The rigidity for a neutral particles is infinitely large, what simply means that neutral particles are not subject to any deflection in the magnetic field.

Appendix B

The Relevant International Conferences

For reference purposes we are listing below the major international conferences relevant to the subject of the book. In the first place we list the International Conferences on Relativistic Nucleus-Nucleus Collisions, called "Quark Matter" conferences, which are the most important conferences in the field of relativistic heavy ion physics. They have been regularly organized since 1979, with intervals of about $1\frac{1}{2}$ years, and in recent years they gather 400–500 participants. Proceedings from these conferences, published as separate volumes of Nuclear Physics A or Journal of Physics G (and once in Zeitschrift für Physik), constitute the richest source of relevant experimental and theoretical information, and are often quoted in this book. We list the "Quark Matter" conferences in Table B.1.

In Table B.2 we list conferences called "Strangeness in Quark Matter", or "SQM". These are smaller conferences, gathering about 150 participants, regularly organized since 1991, with proceedings, except for the first few ones, published in Journal of Physics G. There was some confusion in numbering conferences of this series, we quote the correct numbering.

In Table B.3 we list "Conferences on Nucleus-Nucleus Collisions" which are, however, only in part devoted to relativistic energies. These are big conferences, organized every three years.

Finally, in Table B.4 we list the International Conferences on High Energy Physics and Nuclear Structure (ICOHEPANS), subsequently named Conferences on Particles and Nuclei (PANIC).

Table B.1 The "Quark Matter" conferences.

Number	Year	Place	Proceedings
1st	1979	Berkeley, USA	LBL Report 8957, unpublished
2nd	1982	Bielefeld, FRG	World Scientific, 1982
3rd	1983	Brookhaven, USA	*Nucl. Phys. A* **418** (1984)
4th	1984	Helsinki, Finland	Lecture Notes in Physics Vol. **221**, Springer, 1984
5th	1986	Asilomar, USA	*Nucl. Phys. A* **461** (1987)
6th	1987	Nordkirchen, FRG	*Zeit. Phys. C* **38** (1988)
7th	1988	Lenox, USA	*Nucl. Phys. A* **498** (1989)
8th	1990	Menton, France	*Nucl. Phys. A* **525** (1991)
9th	1991	Gatlinburg, USA	*Nucl. Phys. A* **544** (1992)
10th	1993	Borlänge, Sweden	*Nucl. Phys. A* **566** (1994)
11th	1995	Monterey, USA	*Nucl. Phys. A* **590** (1995)
12th	1996	Heidelberg, Germany	*Nucl. Phys. A* **610** (1996)
13th	1997	Tsukuba, Japan	*Nucl. Phys. A* **638** (1998)
14th	1999	Torino, Italy	*Nucl. Phys. A* **661** (1999)
15th	2001	Long Island, USA	*Nucl. Phys. A* **698** (2002)
16th	2002	Nantes, France	*Nucl. Phys. A* **715** (2003)
17th	2004	Oakland, USA	*J. Phys. G* **30** No.8 (2004)
18th	2005	Budapest, Hungary	*Nucl. Phys. A* **774** (2006)
19th	2006	Shanghai, China	*J. Phys. G* **34** No.8 (2007)
20th	2008	Jaipur, India	*J. Phys. G*
21th	2009	Knoxville, USA	

It would not be possible to list all other conferences, symposia, and workshops related to this field, which have been quite numerous in the last three decades.

International Seminars on Multiquark Interactions and Quantum Chromodynamics (*A. M. Baldin seminars*) are regularly held at JINR, Dubna, since the early 1970s.

Conferences on Physics and Astrophysics of Quark-Gluon Plasma are organized in India: the first one took place in Mumbai (Bombay) in 1988, and four consecutive ones alternately in Jaipur and Kolkata (Calcutta).

One can also quote "Quark-Gluon Plasma Signatures", "Critical Point and Onset of Deconfinement", "Hot Quarks", and many others.

Also, in major serial conferences on high energy physics: the so-called "Rochester Conferences" organized every two years, and "Europhysics Conferences" organized in alternate years, a special session is usually devoted to collisions of relativistic nuclei, and to quark-gluon plasma.

Table B.2 The "Strangeness in Quark Matter" conferences.

Number	Year	Place	Proceedings
1st	1991	Aarhus, The Netherlands	*Nucl. Phys. B (Proc. Suppl.)* **24** (1991)
2nd	1994	Kolymbari (Krete), Greece	World Scientific 1995
3rd	1995	Tucson, USA	*AIP Conf. Proc.* **340** (1995)
4th	1996	Budapest, Hungary	*Heavy Ion Physics* **4** (1996)
5th	1997	Thera (Santorini), Greece	*J. Phys. G* **23** No.12 (1997)
6th	1998	Padova, Italy	*J. Phys. G* **25** No.2 (1999)
7th	2000	Berkeley, USA	*J. Phys. G* **27** No.3 (2001)
8th	2001	Frankfurt, Germany	*J. Phys. G* **28** No.7 (2002)
9th	2003	Atlantic Beach, USA	*J. Phys. G* **30** No.1 (2004)
10th	2004	Cape Town, RSA	*J. Phys. G* **31** No.6 (2005)
11th	2006	Los Angeles, USA	*J. Phys. G* **32** No.12 (2006)
12th	2007	Levoča, Slovakia	*J. Phys. G* **35** No.4 (2008)
13th	2008	Beijing, China	*J. Phys. G*
14th	2009	Buzios, Brazil	

Table B.3 International Conferences on Nucleus-Nucleus Collisions.

Number	Year	Place	Proceedings
1st	1982	East Lansing, USA	*Nucl. Phys. A* **400** (1983)
2nd	1985	Visby, Sweden	*Nucl. Phys. A* **447** (1986)
3rd	1988	Saint Malo, France	*Nucl. Phys. A* **488** (1988)
4th	1991	Kanazawa, Japan	*Nucl. Phys. A* **538** (1992)
5th	1994	Taormina, Italy	*Nucl. Phys. A* **583** (1995)
6th	1997	Gatlinburg, USA	*Nucl. Phys. A* **630** (1998)
7th	2000	Strasbourg, France	*Nucl. Phys. A* **685** (2001)
8th	2003	Moscow, Russia	*Nucl. Phys. A* **734** (2004)
9th	2006	Rio de Janeiro, Brazil	*Nucl. Phys. A* **787** (2007)
10th	2009	Beijing, China	

Table B.4 International Conferences on Particles and Nuclei (listed from 1977).

Number	Year	Place	Proceedings
7th	1977	Zürich, Switzerland	Birkhäuser Verlag 1977
8th	1979	Vancouver, Canada	*Nucl. Phys. A* **435** (1980)
9th	1981	Versailles, France	*Nucl. Phys. A* **374** (1982)
10th	1984	Heidelberg, Germany	*Nucl. Phys. A* **434** (1985)
11th	1987	Kyoto, Japan	*Nucl. Phys. A* **478** (1988)
12th	1990	Cambridge, USA	*Nucl. Phys. A* **527** (1991)
13th	1993	Perugia, Italy	
14th	1996	Williamsburg, USA	World Scientific 1997
15th	1999	Uppsala, Sweden	*Nucl. Phys. A* **663–664** (2000)
16th	2002	Osaka, Japan	*Nucl. Phys. A* **721** (2003)
17th	2005	Santa Fe, USA	*AIP Conf. Proc.* **842** (2006)
18th	2008	Eilat, Israel	*Nucl. Phys. A*

Index